SpringerBriefs in Physics

SpringerBriefs in Physics are a series of slim high-quality publications encompassing the entire spectrum of physics. Manuscripts for SpringerBriefs in Physics will be evaluated by Springer and by members of the Editorial Board. Proposals and other communication should be sent to your Publishing Editors at Springer.

Featuring compact volumes of 50 to 125 pages (approximately 20,000–45,000 words), Briefs are shorter than a conventional book but longer than a journal article. Thus, Briefs serve as timely, concise tools for students, researchers, and professionals.

Typical texts for publication might include:

- A snapshot review of the current state of a hot or emerging field
- A concise introduction to core concepts that students must understand in order to make independent contributions
- An extended research report giving more details and discussion than is possible in a conventional journal article
- A manual describing underlying principles and best practices for an experimental technique
- An essay exploring new ideas within physics, related philosophical issues, or broader topics such as science and society

Briefs allow authors to present their ideas and readers to absorb them with minimal time investment.

Briefs will be published as part of Springer's eBook collection, with millions of users worldwide. In addition, they will be available, just like other books, for individual print and electronic purchase.

Briefs are characterized by fast, global electronic dissemination, straightforward publishing agreements, easy-to-use manuscript preparation and formatting guidelines, and expedited production schedules. We aim for publication 8–12 weeks after acceptance.

More information about this series at http://www.springer.com/series/8902

Valeriya Akhmedova · Emil T. Akhmedov

Selected Special Functions for Fundamental Physics

 Springer

Valeriya Akhmedova
Kharkevich Institute for Information
Transmission Problems
Moscow, Russia

HSE University
Moscow, Russia

Emil T. Akhmedov
Moscow Institute of Physics
and Technology
Moscow, Russia

Institute for Theoretical
and Experimental Physics
Moscow, Russia

ISSN 2191-5423   ISSN 2191-5431   (electronic)
SpringerBriefs in Physics
ISBN 978-3-030-35088-8   ISBN 978-3-030-35089-5   (eBook)
https://doi.org/10.1007/978-3-030-35089-5

This Springer imprint is published by the registered company Springer Nature Switzerland AG
The registered company address is: Gewerbestrasse 11, 6330 Cham, Switzerland

Preface

These notes are based on the seminar course given at the Moscow Institute of Physics and Technology. We present calculational methods which are used both in mathematical and theoretical physics. That includes application of the advanced complex analysis in fundamental physics—from quantum mechanics to quantum field theory in curved space–time.

Moscow, Russia

Valeriya Akhmedova
Emil T. Akhmedov

Acknowledgements

We would like to thank Hermann Nicolai and Stefan Theisen for the hospitality at the Albert Einstein Institute, Golm, where the work on this project was completed. The work of ETA was supported by the state grant Goszadanie 3.9904.2017/BCh and by the grant from the Foundation for the Advancement of Theoretical Physics and Mathematics "BASIS" and by RFBR grant 18-01-00460A.

Contents

1   **Introduction**   1

2   **Γ-Function**   3
    2.1   Some Properties   4
    2.2   Weierstrass Representation   8
    2.3   Stirling Formula   9
    2.4   Contour Integral Representation   11
    2.5   Euler's B-Function   12

3   **Riemann ζ-Function**   15
    3.1   Integral Representation   15
    3.2   Euler's Infinite Product   17
    3.3   Riemann's Hypothesis   18
    3.4   Application: Functional Determinant   19

4   **Hermite Polynomials**   23
    4.1   Application: Schrödinger Equation   23
    4.2   Definition   26
    4.3   Generating Function   27
    4.4   Recurrence Relations   28
    4.5   Integral Representation   29
    4.6   Fourier Transformation   29
    4.7   Orthogonality   30
    4.8   Asymptotic Form for the Large Index   32
    4.9   Completeness   33
    4.10   Relation to the Representations of the Heisenberg Algebra   35
    4.11   Applications: Back to the Quantum Oscillator   37

5   **Bessel Functions**   41
    5.1   Generating Function   42
    5.2   Series Expansion   43

5.3   Bessel Function $J_v(z)$ with Complex Index $v \in \mathbb{C}$ 44
5.4   Recurrence Relations for $J_v(z)$ 44
5.5   Bessel Function of the Second Kind 45
5.6   Series Expansion for $Y_m(z)$ 46
5.7   Hankel and MacDonald Functions 47
5.8   Bessel, Hankel and MacDonald Functions of Half-Integer
        Indexes 48
5.9   Integral Representation 49
5.10   Asymptotic Form for the Large Argument 52
5.11   Orthogonality 53
5.12   Addition or Summation Theorems for $J_m(z)$ 54
5.13   Relation to the Group Representation Theory 55
5.14   Application: General Discussion of the Green Functions 57
5.15   Application: Green Functions of the Klein–Gordon Equation . . . 61

6 **Legendre Polynomials and Spherical Functions** 65
6.1   Generating Function and Integral Representation 66
6.2   Recurrence Relations 68
6.3   Orthogonality 69
6.4   Asymptotic Form for the Large Index 71
6.5   Completeness 72
6.6   Spherical Harmonics 72
6.7   Relation to the Representation Theory 73
6.8   Integral Representation of $P_n^m(\cos\theta)$ 75
6.9   Addition or Summation Theorems 77
6.10   Legendre Functions $P_v^\mu(z)$ for $\mu, v \in \mathbb{C}$ 78

7 **Hypergeometric Functions** 81
7.1   Behavior in the Vicinities of the Peculiar Points 83
7.2   Hypergeometric Series 84
7.3   Integral Representation and Analytical Continuation 86
7.4   Contour Barnes Integral Representation 86
7.5   Elementary Properties 88
7.6   Functional Relations Between Hypergeometric Functions 90
7.7   Asymptotic Form for the Large Argument 93
7.8   Relation to the Legendre Functions 94
7.9   Application: The Feynman Propagator on the Sphere 95

8 **Degenerate Hypergeometric Function** 99
8.1   Differential Equation 99
8.2   Integral Representation 100
8.3   Laplace Transformation 101
8.4   Another Integral Representation from the Laplace
        Transformation 102

8.5   Simplest Relations 103
8.6   Asymptotic Behavior for the Large Argument 104
8.7   Relations to Other Functions 104

9 θ-Functions 107
9.1   Different Types of θ-Functions 108
9.2   Zeros of the θ-Functions 109
9.3   Composition Equations 111
9.4   Infinite Product Representation 112

Chapter 1
Introduction

There are many excellent books on the subject of special functions and their applications in physics. In this sense our course cannot add much to the subject. However, our main goal was to present the methods which allow one to work with special functions and more generally with differential equations and to apply the advanced complex analysis in the fundamental physics context. We restrict our attention to some special functions, which are the most frequently used in quantum mechanics, theory of relativity and quantum field theory.

In some places our presentation may sound naive from the rigorous mathematical point of view. However, one should realize that we avoid providing rigorous proofs of various theorems unless that is necessary to develop the constructive calculational methods. E.g., if in the notes we exchange an integration with a summation without explicitly stating that this is possible it means that it is straightforward to show that the sum and the integral are convergent. Or if we perform an analytical continuation it means that we assume it is straightforward to see that the expression under consideration is analytic in the appropriate region of the complex plane. Moreover, we prefer to show how the methods work on concrete examples instead of proving general theorems.

We have assumed that students who were attending this course were familiar with the basics of the calculus, linear algebra, complex analysis and with the basic methods of solving differential equations and of calculating integrals.

The books that have been used in writing these notes are as follows:

- Harry Hochstadt, "The functions of mathematical physics (Dover books on physics)", 2012.
- N. N. Lebedev, "Special functions and their applications (Dover books on mathematics)", 1972.
- E. T. Whittaker and G. N. Watson, "A Course of Modern Analysis", 1927.

© The Author(s), under exclusive license to Springer Nature Switzerland AG 2019
V. Akhmedova and E. T. Akhmedov, *Selected Special Functions*
for Fundamental Physics, SpringerBriefs in Physics,
https://doi.org/10.1007/978-3-030-35089-5_1

- L. Landau and E. Lifshitz, III-rd volume, Quantum mechanics, Course of theoretical physics.
- S. Khoroshkin, unpublished notes of the lectures presented in the mathematical faculty of the HSE, Moscow.

Chapter 2
Γ-Function

Abstract This section is recorded by MIPT students Petrova Elena and Ivanenko Aleksei. It is about the properties of the Γ-function, which are used in the other sections of this book.

The definition of the Γ-function is as follows:

$$\Gamma(z) = \int_0^\infty e^{-t} t^{z-1} \, dt, \tag{2.1}$$

where $z \in \mathbb{C}$. For Re $z > 0$ the integral converges at the lower limit of integration.

The properties of the Γ-function that follow immediately from its definition are as follows. The first one is:

$$\Gamma(n+1) = n!, \quad n \in \mathbb{N}.$$

This property can be shown by the direct integration by part. The second property is that:

$$\Gamma\left(\frac{1}{2}\right) = \int_0^\infty e^{-t} t^{-\frac{1}{2}} \, dt = 2 \int_0^\infty e^{-u^2} \, du = \sqrt{\pi}.$$

Now let us continue with the analytic properties of the Γ-function in the complex z-plane. Divide the integration in (2.1) as follows:

$$\Gamma(z) = \int_0^1 e^{-t} t^{z-1} \, dt + \int_1^\infty e^{-t} t^{z-1} \, dt \equiv P(z) + Q(z).$$

The integral for $P(z)$ converges absolutely and homogeneously for Re $z > 0$. At the same time the integral for $Q(z)$ converges absolutely and homogeneously for Re $z < \infty$. Hence, $P(z)$ and $Q(z)$ are analytic in those regions of the complex z-plane.

© The Author(s), under exclusive license to Springer Nature Switzerland AG 2019
V. Akhmedova and E. T. Akhmedov, *Selected Special Functions for Fundamental Physics*, SpringerBriefs in Physics,
https://doi.org/10.1007/978-3-030-35089-5_2

Furthermore,

$$P(z) = \int_0^1 t^{z-1} \sum_{k=0}^{\infty} \frac{(-1)^k t^k}{k!} \, dt = \sum_{k=0}^{\infty} \frac{(-1)^k}{k!} \int_0^1 t^{k+z-1} \, dt = \sum_{k=0}^{\infty} \frac{(-1)^k}{k!} \frac{1}{z+k}.$$

Thus,

$$\Gamma(z) = \sum_{k=0}^{\infty} \frac{(-1)^k}{k!} \frac{1}{z+k} + \int_1^{\infty} e^{-t} t^{z-1} \, dt.$$

In the vicinity of $z = -n$, where $n \in \mathbb{N}$ we have that

$$\Gamma(z) = \frac{(-1)^n}{n!} \frac{1}{z+n} + F(z),$$

where $F(z)$ is a regular at $z = -n$ function. Hence, $\Gamma(z)$ has poles at $z = 0, -1,$ $-2, \ldots$ with the residues $\text{Res}_{z=-n} \Gamma(z) = \frac{(-1)^n}{n!}$.

2.1 Some Properties

One of the simplest relations which Γ-function does obey is as follows:

$$\boxed{\Gamma(z+1) = z\,\Gamma(z), \quad \text{Re}\, z > 0.}$$

In fact, let us preform the integration by parts:

$$\Gamma(z+1) = \int_0^{\infty} e^{-t} t^z \, dt = -e^{-t} t^z \Big|_0^{\infty} + z \int_0^{\infty} e^{-t} t^{z-1} \, dt = z\,\Gamma(z), \quad z \neq 0, -1, -2, \ldots.$$

via the analytical continuation this can be shown to be valid for all values of z, except the poles.

From the already obtained relations it immediately follows that:

$$\Gamma\left(n + \frac{1}{2}\right) = \frac{1 \cdot 3 \cdot 5 \ldots (2n-1)}{2^n} \sqrt{\pi}, \quad \text{for} \quad n \in \mathbb{N}.$$

Let us prove the relation as follows:

$$\boxed{\Gamma(z)\,\Gamma(1-z) = \frac{\pi}{\sin(\pi z)}.} \tag{2.2}$$

This relation can be shown in the following way: both sides of it are analytic in the entire complex z-plane (including infinity) and have the same poles and residues in them. Hence, the difference of these two functions is a constant according to the Cauchy theorem. The constant can be calculated at any regular point in the complex z-plane and can be shown to be zero. This proves the relation in question.

The above reasoning is the standard and frequently used method to prove relations between complex functions,[1] but to show various methods below we also will provide other ways to establish the same relations. Namely, another way to see the relation (2.2) is as follows: Let $0 < \mathrm{Re}\, z < 1$, then:

$$
\Gamma(z)\,\Gamma(1 - z) = \int_0^\infty ds \int_0^\infty dt\, e^{-(s+t)} s^{-z} t^{z-1}.
$$

Make the following change of variables in the last integral: $u = s + t$, $v = \frac{t}{s}$. Then

$$
\Gamma(z)\,\Gamma(1 - z) = \int_0^\infty \int_0^\infty e^{-u} v^{z-1} \frac{du\,dv}{1 + v} = \int_0^\infty \frac{v^{z-1}}{1 + v}\,dv.
$$

Let us show that

$$
\int_0^\infty \frac{v^{z-1}}{1 + v}\,dv = \frac{\pi}{\sin(\pi z)}.
$$

Consider the integral:

$$
\int_{C'} \frac{(-v)^{z-1}}{1 + v}\,dv,
$$

where the contour $C' = C_1 + C_2 + C_3$ is such as shown on the figure:

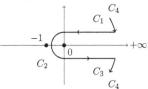

C_4 here is the large circle.

---

[1]E.g. in the case of the previous relation one could compare the analytic properties of $1/z$ and $\Gamma(z)/\Gamma(z + 1)$ functions. Note that one cannot do the same for $\Gamma(z + 1)$ and $z\,\Gamma(z)$, because these functions are not analytic at $z = \infty$, as we will see below in the Sect. 2.3. Note that $e^{1/z}$ is not analytic at $z = 0$. Similarly e^z is not analytic at infinity, because the proper coordinate at infinity is $w = 1/z$.

Then,

$$\int_{C'} \frac{(-v)^{z-1}}{1+v}\, dv = \int_{C'} \frac{e^{(z-1)[\log|v|+i(\arg v-\pi)]}}{v+1}\, dv =$$

$$= e^{-i\pi(z-1)} \int_{C_1} \frac{v^{z-1}}{1+v}\, dv + e^{i\pi(z-1)} \int_{C_3} \frac{v^{z-1}}{1+v}\, dv +$$

$$+ \lim_{r\to 0} \int_{-\pi}^{\pi} r^{z-1}\, e^{i(z-1)\varphi}\, i\, r\, e^{i\varphi}\, \frac{d\varphi}{1+re^{i\varphi}}.$$

The last integral corresponds to the half-circular contour C_2, on which $v = re^{i(\varphi+\pi)}$, where r is the radius of this small half circle. Then,

$$\int_{C'} \frac{(-v)^{z-1}}{v+1}\, dv = e^{-i\pi z} \int_0^\infty \frac{v^{z-1}}{v+1}\, dv - e^{i\pi z} \int_0^\infty \frac{v^{z-1}}{v+1}\, dv + 0.$$

The last contribution is 0, because it is $\sim r^{\operatorname{Re} z}$, as $r \to 0$. Note that we have assumed above that Re $z > 0$.

On the big circular contour C_4 one can put $v = R\, e^{i(\varphi+\pi)}$, where R is the radius of the circle, and

$$\int_{C_4} \frac{(-v)^{z-1}}{v+1}\, dv = \lim_{R\to\infty} \int_{\pi}^{-\pi} \frac{R^z}{1-R\,e^{i\varphi}}\, e^{iz\varphi}\, d\varphi.$$

The value of this integral tends to zero as $R^{\operatorname{Re} z-1}$, when $R \to \infty$. Note that we have assumed above that Re $z < 1$. Thus, combining all the above relations, one can see that:

$$\left(e^{-i\pi z} - e^{i\pi z}\right) \int_0^\infty \frac{v^{z-1}}{v+1}\, dv = \oint_C \frac{-v^{z-1}}{v+1}\, dv = -2\pi i.$$

where $C = C' + C_4$—is the clockwise contour around the single pole of the integrand at $v = -1$.

Hence, we obtain that:

$$\boxed{\int_0^\infty \frac{v^{z-1}}{1+v}\, dv = \frac{\pi}{\sin(\pi z)},}$$

which concludes the proof of Eq. (2.2). Via analytical continuation we can extend this relation beyond the stripe $0 < \operatorname{Re} z < 1$.

Let us continue with the proof of the following relation:

$$2^{2z-1} \Gamma(z) \Gamma\left(z + \frac{1}{2}\right) = \Gamma\left(\frac{1}{2}\right) \Gamma(2z) = \sqrt{\pi} \, \Gamma(2z). \qquad (2.3)$$

To prove this property consider the product of integrals defining $\Gamma(z)$ and $\Gamma(z + 1/2)$:

$$2^{2z-1} \Gamma(z) \Gamma\left(z + \frac{1}{2}\right) =$$

$$= \int_0^\infty \int_0^\infty e^{-(s+t)} (2\sqrt{s\,t})^{2z-1} t^{-\frac{1}{2}} \, ds \, dt =$$

$$= 4 \int_0^\infty \int_0^\infty e^{-(\alpha^2+\beta^2)} (2\alpha\beta)^{2z-1} \, \alpha \, d\alpha \, d\beta.$$

To obtain the last line we have changed the variables as $\sqrt{s} = \alpha$, $\sqrt{t} = \beta$. If we exchange in the obtained integral α and β with each other and add up halves of the resulting expressions to each other, we find that:

$$2^{2z-1} \Gamma(z) \Gamma\left(z + \frac{1}{2}\right) = 2 \int_0^\infty \int_0^\infty e^{-(\alpha^2+\beta^2)} (2\alpha\beta)^{2z-1} (\alpha+\beta) \, d\alpha \, d\beta.$$

This expression is symmetric under the exchange of α and β. Now the last integral can be rewritten as

$$2^{2z-1} \Gamma(z) \Gamma\left(z + \frac{1}{2}\right) = 4 \iint_M d\alpha \, d\beta \, e^{-(\alpha^2+\beta^2)} (2\alpha\beta)^{2z-1} (\alpha+\beta).$$

where M is the region in which: $0 \leqslant \alpha \leqslant \infty$, $0 \leqslant \beta \leqslant \alpha$. Then, changing the variables as $u = \alpha^2 + \beta^2$ and $v = 2\alpha\beta$, we obtain:

$$2^{2z-1} \Gamma(z) \Gamma\left(z + \frac{1}{2}\right) = \int_0^\infty v^{2z-1} \, dv \int_0^\infty \frac{e^{-u}}{\sqrt{u-v}} \, du$$

$$= 2 \int_0^\infty e^{-v} v^{2z-1} \, dv \int_0^\infty e^{-w^2} \, dw = \sqrt{\pi} \, \Gamma(2z),$$

which concludes the proof of the relation (2.3).

2.2 Weierstrass Representation

The Weierstrass representation of the Γ-function is:

$$\frac{1}{\Gamma(z+1)} = e^{\gamma z} \prod_{n=1}^{\infty} e^{-\frac{z}{n}} \left(1 + \frac{z}{n}\right). \tag{2.4}$$

Here $\psi(z) = \frac{\Gamma'(z)}{\Gamma(z)}$ and $\psi(1) = \Gamma'(1) = -\gamma = -0,5772156...$—is the so called Euler constant.

To prove the Weierstrass representation consider the following expression:

$$\Gamma'(z) = \int_{0}^{\infty} e^{-t} t^{z-1} \log t \, dt, \quad \text{for} \quad \text{Re} \, z > 0.$$

Let us use in this expression the following relation $\log t = \int_{0}^{\infty} \frac{e^{-x} - e^{-xt}}{x} dx$, which is valid for Re $t > 0$. (This representation for the logarithm can be proved by the integration of both sides of the relation $\frac{1}{t'} = \int_{0}^{+\infty} e^{-xt'} dx$ from $t' = 1$ to $t' = t$.)
Then,

$$\Gamma'(x) = \int_{0}^{\infty} \frac{dx}{x} \int_{0}^{\infty} [e^{-x} - e^{-xt}] e^{-t} t^{z-1} dt = \int_{0}^{\infty} \frac{dx}{x} \left[e^{-x} \Gamma(z) - \int_{0}^{\infty} e^{-t(x+1)} t^{z-1} dt \right]. \tag{2.5}$$

If in the last integral over dt we make the following change of variables $u = t(x + 1)$, then it becomes apparent that it is equal to $(x + 1)^{-z} \Gamma(z)$. Hence, from (2.5) we obtain that

$$\psi(z) \equiv \frac{\Gamma'(z)}{\Gamma(z)} = \int_{0}^{\infty} \left[e^{-x} - \frac{1}{(x+1)^z} \right] \frac{dx}{x},$$

for Re $z > 0$. Let us now, using the change of variables $x \to t$ in the first integral and $x + 1 = e^t$ in the second integral, represent the last expression as:

$$\psi(z) = \lim_{\delta \to 0} \left[\int_{\delta}^{\infty} \frac{e^{-t}}{t} dt - \int_{\log(1+\delta)}^{\infty} \frac{e^{-tz}}{1 - e^{-t}} dt \right]$$

$$= \lim_{\delta \to 0} \left[\int_{\log(1+\delta)}^{\infty} \left[\frac{e^{-t}}{t} - \frac{e^{-tz}}{1 - e^{-t}} \right] dt - \int_{\log(1+\delta)}^{\delta} \frac{e^{-t}}{t} dt \right].$$

Then, we have that

$$\lim_{\delta \to 0} \int_{\log(1+\delta)}^{\delta} \frac{e^{-t}}{t} dt = 0.$$

Hence,

$$\psi(z) = \int_0^\infty \left[\frac{e^{-t}}{t} - \frac{e^{-tz}}{1 - e^{-t}} \right] dt, \quad \text{where} \quad \text{Re } z > 0. \tag{2.6}$$

Let us put in this expression $z = 1$ and then subtract the obtained relation from the both sides of (2.6). As the result, we obtain that:

$$\psi(z) = -\gamma + \int_0^\infty \frac{e^{-t} - e^{-tz}}{1 - e^{-t}} dt,$$

where $\gamma = -\psi(1)$ is the defined above Euler's constant. If we change $x = e^{-t}$ in the last expression, then we find that:

$$\psi(z) = -\gamma + \int_0^1 \frac{1 - x^{z-1}}{1 - x} dx.$$

If in this expression one Taylor expands $\frac{1}{1-x}$ and takes the integral in each member of the found this way series, then he obtains the following relation:

$$\boxed{\psi(z) = -\gamma + \sum_{n=0}^\infty \left[\frac{1}{n+1} - \frac{1}{n+z} \right].}$$

Finally, after the change $z \to z + 1$ and the integration of this expression from 0 to z over an arbitrary contour, we obtain the Weierstrass relation (2.4).

2.3 Stirling Formula

Consider the function

$$\Gamma(z+1) = \int_0^\infty e^{-t} t^z dt = \int_0^\infty e^{-(t - z \ln t)} dt.$$

We would like to estimate this integral as Re $z \to \infty$ by the steepest descent or the stationary phase approximation method. This method works as follows. Consider the following limit of the integral:

$$\lim_{\lambda \to \infty} \int_a^b f(t) e^{\lambda s(t)} dt \approx \sum_q \int_a^b f(t_q) e^{\lambda s(t_q) + \lambda s''(t_q)(t - t_q)^2} dt \approx$$

$$\approx \sum_q \int_{-\infty}^{+\infty} f(t_q) e^{\lambda s(t_q) + \lambda s''(t_q)(t - t_q)^2} dt \approx \sum_q \sqrt{\frac{\pi}{-\lambda s''(t_q)}} e^{\lambda s(t_q)} f(t_q),$$

where $s'(t_q) = 0$, i.e. t_q are the extrema of the function $s(t)$ on the interval (a, b), index q enumerates all of them and we assume that $f(t)$ is some regular function on the interval. We have extended the limits of integration from $[a, b]$ to $(-\infty, +\infty)$ because the integral is rapidly convergent as $\lambda \to \infty$. Taylor expanding $f(t)$ and $s(t)$ around t_q's one can show that the corrections to the last expression are suppressed by the powers of $1/\lambda$.

In our case we can define $s(t) \equiv (t - z \ln t) - (z - z \ln z)$. Then,

$$s'(t) = 1 - \frac{z}{t}, \quad \text{and} \quad s''(t) = \frac{z}{t^2}.$$

Hence, $s(z) = s'(z) = 0$ and $s''(z) = \frac{1}{z}$.

As the result, the Taylor expansion of $s(t)$ around the extremum $t = z$ is as follows:

$$s(t) = \frac{1}{2z}(t - z)^2 + \dots .$$

Then as Re $z \to +\infty$, we have that

$$\Gamma(z + 1) \approx e^{-(z - z \ln z)} \int_{-\infty}^{+\infty} e^{-\frac{1}{2z}(t - z)^2} dt \approx \sqrt{2\pi z} \left(\frac{z}{e}\right)^z .$$

Again the limits of integration over t are extended here to $\pm\infty$ because the integral is rapidly converging. Hence, we obtain that

$$\boxed{\Gamma(z) \approx \sqrt{\frac{2\pi}{z}} \left(\frac{z}{e}\right)^z, \quad \text{as} \quad z \to +\infty,}$$

which is the so called Stirling formula.

2.4 Contour Integral Representation

Consider the following integral

$$F(z) = \int_C e^{-\xi} \xi^{z-1} d\xi,$$

where the contour of integration $C = I + II + C_r$ is shown on the figure:

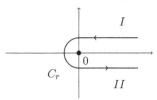

One can split the integral under consideration as $F(z) = \int\limits_I + \int\limits_{C_r} + \int\limits_{II}$. Represent in it $\xi^{z-1} = e^{(z-1)\log\xi}$, where $\log \xi$ is taken on such a sheet of the complex ξ-plane that $0 < \arg \xi \leq 2\pi$. On I part of the contour we have that $\xi = t$, $t \in \mathbb{R}$, while on II one can represent $\xi = t\, e^{2\pi i}$. Then,

$$F(z) = (e^{2\pi i z} - 1) \int_r^\infty e^{-t}\, t^{z-1}\, dt + \int_{C_r} e^{-\xi} \xi^{z-1}\, d\xi.$$

On C_r one can represent $\xi = r\, e^{i\varphi}$. Hence, on C_r we have that $\left| e^{-\xi} \xi^{z-1} \right| = e^{-r\cos\varphi} e^{(x-1)\log r - \varphi y} < A\, r^{x-1}$, for some real constant A. Here we have used the decomposition $z = x + iy$, where $x, y \in \mathbb{R}$.

Then, for Re $z = x > 0$ one can see that the integral over the half-circle $\int\limits_{C_r}$ is vanishing, as $r \to 0$, and

$$F(z) = (e^{2\pi i z} - 1) \int_0^\infty e^{-t} t^{z-1} dt = \left(e^{2\pi i z} - 1\right) \Gamma(z).$$

Hence,

$$\Gamma(z) = \frac{1}{e^{2\pi i z} - 1} \int_C e^{-\xi} \xi^{z-1} d\xi.$$

Changing in this equation $z \to 1 - z$, we obtain:

$$\Gamma(1-z) = \frac{1}{e^{-2\pi i z} - 1} \int_C e^{-\xi} \xi^{-z} d\xi = \frac{e^{-\pi i z}}{e^{-2\pi i z} - 1} \int_C e^{-\xi} (-\xi)^{-z} d\xi$$

$$= \frac{i}{2 \sin \pi z} \int_C e^{-\xi} (-\xi)^{-z} d\xi$$

Using here Eq. (2.2), we find the contour integral representation as follows:

$$\boxed{\frac{1}{\Gamma(z)} = \frac{1}{2\pi i} \int_{C*} e^{\xi} \xi^{-z} d\xi,} \qquad (2.7)$$

where the contour of integration C^* is shown on the figure:

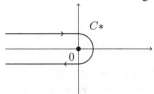

Note that to obtain (2.7) we have made the change $\xi \rightarrow -\xi$.

2.5 Euler's B-Function

Consider the Euler's B-function

$$\boxed{B(x, y) \equiv \int_0^1 t^{x-1}(1-t)^{y-1} dt,}$$

where Re $x > 0$ and Re $y > 0$.

   Make in it the following change of variables $u = \frac{t}{1-t}$. Then, we obtain that:

$$B(x, y) = \int_0^\infty \frac{u^{x-1}}{(1+u)^{x+y}} du.$$

Now take into account that

$$\int_0^\infty e^{-pt} t^{z-1} dt = \frac{\Gamma(z)}{p^z}.$$

Let us represent here $p = 1 + u$ and $z = x + y$. Then, we obtain:

$$\frac{1}{(1+u)^{x+y}} = \frac{1}{\Gamma(x+y)} \int_0^\infty e^{-(1+u)t} t^{x+y-1} \, dt.$$

Substituting this expression into the last representation of $B(x, y)$, we find the following relation:

$$B(x, y) = \frac{1}{\Gamma(x+y)} \int_0^\infty e^{-t} t^{x+y-1} \, dt \int_0^\infty e^{-ut} u^{x-1} \, du = \frac{\Gamma(x)}{\Gamma(x+y)} \int_0^\infty e^{-t} t^{y-1} \, dt.$$

And, hence, it follows that

$$\boxed{B(x, y) = \frac{\Gamma(x)\,\Gamma(y)}{\Gamma(x+y)},}$$

which is the well known representation of the Euler's B-function.

Chapter 3
Riemann ζ-Function

Abstract This section is about the properties of the Riemann ζ-function. Here we also discuss Riemann hypothesis and the uses of the ζ-function in the calculations of functional integrals.

Let Re $z > 1$. Then, the following series is convergent:

$$\zeta(z) \equiv \sum_{n=1}^{+\infty} \frac{1}{n^z}.$$

Hence, in the region Re $z > 1$ this is an analytic function of z. It is referred to as the Riemann ζ-function.

We will also consider the generalized ζ-function:

$$\zeta(z, a) \equiv \sum_{n=0}^{+\infty} \frac{1}{(n+a)^z},$$

and we assume that $0 < a \leq 1$. Obviously $\zeta(z, 1) = \zeta(z)$.

3.1 Integral Representation

From the section on Γ-functions we know that:

$$(a + n)^{-z} \, \Gamma(z) = \int_0^{+\infty} x^{z-1} \, e^{-(n+a)x} \, dx,$$

when Re $z > 0$. Then,

$$\Gamma(z) \, \zeta(z, a) = \sum_{n=0}^{+\infty} \int_0^{+\infty} x^{z-1} \, e^{-(n+a)x} \, dx = \int_0^{+\infty} \frac{x^{z-1} \, e^{-ax}}{1 - e^{-x}} \, dx,$$

© The Author(s), under exclusive license to Springer Nature Switzerland AG 2019
V. Akhmedova and E. T. Akhmedov, *Selected Special Functions*
for Fundamental Physics, SpringerBriefs in Physics,
https://doi.org/10.1007/978-3-030-35089-5_3

and, hence,

$$\zeta(z, a) = \frac{1}{\Gamma(z)} \int_0^{+\infty} \frac{x^{z-1} e^{-ax}}{1 - e^{-x}} \, dx.$$

Consider the following integral:

$$\int_C \frac{(-\xi)^{z-1} e^{-a\xi}}{1 - e^{-\xi}} \, d\xi,$$

where Re $z > 1$ and we deal here with the complex ξ-plane with the cut $|\arg(-\xi)| \le \pi$. The contour C in this complex plane is defined as shown on the figure:

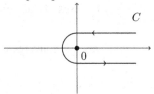

We assume that the contour C does not pass through the points $\xi = 2\pi i n$, $n \in \mathbb{Z}$. The latter are the poles of the integrand under consideration. Performing the same manipulations as in the section on Γ-functions, one can show that the integral under consideration is equal to:

$$\int_C \frac{(-\xi)^{z-1} e^{-a\xi}}{1 - e^{-\xi}} \, d\xi = \left[e^{\pi i (z-1)} - e^{-\pi i (z-1)} \right] \int_0^{+\infty} \frac{x^{z-1} e^{-ax}}{1 - e^{-x}} \, dx.$$

Hence, using the relation (2.2), one can establish that:

$$\zeta(z, a) = -\frac{\Gamma(1-z)}{2\pi i} \int_C \frac{(-\xi)^{z-1} e^{-a\xi}}{1 - e^{-\xi}} \, d\xi. \tag{3.1}$$

The integral here is an analytic function for all values of z. Because it is convergent for all those values. Hence, all peculiar points of $\zeta(z, a)$ in the complex z-plane can be at most those of $\Gamma(1 - z)$, which are are $z = 1, 2, 3, \ldots$, i.e. $\zeta(z, a)$ is analytic in the complex z-plane except may be those points. But we have seen above that $\zeta(z, a)$ is analytic for Re $z > 1$. Hence, the only peculiar point of $\zeta(z, a)$ is at $z = 1$. The other poles of $\Gamma(1 - z)$ should be compensated by the zeros of the integral in (3.1).

If we put $z = 1$ in the integral (3.1), then we obtain:

$$\frac{1}{2\pi i} \int_C \frac{e^{-a\xi}}{1 - e^{-\xi}} \, d\xi = 1,$$

because the integral here is equal to the only residue of the integrand at $\xi = 0$ outside the contour C. Thus,

$$\zeta(z, a) = \frac{1}{z-1} + \dots, \quad \text{as} \quad z \to 1,$$

which is the only pole of the function $\zeta(z, a)$ in the complex z-plane.

3.2 Euler's Infinite Product

Subtract from the series defining $\zeta(z)$ the series for $2^{-z} \zeta(z)$, then we obtain:

$$\zeta(z) \left(1 - 2^{-z}\right) = \frac{1}{1^z} + \frac{1}{3^z} + \frac{1}{5^z} + \dots = \sum_{n=0}^{+\infty} \frac{1}{(2n+1)^z}.$$

Here the summation goes only over odd numbers. Similarly:

$$\zeta(z) \left(1 - 2^{-z}\right) \left(1 - 3^{-z}\right) = \frac{1}{1^z} + \frac{1}{5^z} + \frac{1}{7^z} + \dots.$$

In the latter series all the terms in which n divides 2 and 3 are absent.
Furthermore,

$$\zeta(z) \left(1 - 2^{-z}\right) \left(1 - 3^{-z}\right) \dots \left(1 - p^{-z}\right) = 1 + \sum{}' n^{-z},$$

where the product on the left hand side (LHS) goes over all prime numbers up to p and the sum $\sum'$ on the right hand side (RHS) is taken over such n that are grater than p and do not divide any prime number smaller than p. But

$$\left| \sum{}' n^{-z} \right| \leq \sum{}' n^{-\operatorname{Re} z} \leq \sum_{n=p+1}^{+\infty} n^{-\operatorname{Re} z} \to 0, \quad \text{as} \quad p \to \infty.$$

Thus, for $\operatorname{Re} z > 1$, the infinite product $\zeta(z) \prod_p (1 - p^{-z})$, where p runs over all primes, is convergent and is equal to one. Hence, we obtain:

$$\boxed{\zeta(z) = \frac{1}{\prod_p \left(1 - \frac{1}{p^z}\right)}, \quad \text{for} \quad \operatorname{Re} z > 1,}$$

which is the so called Euler's product formula for the ζ-function.
    This infinite product formula for the Riemann ζ-function has a very simple heuristic explanation. If one Taylor expands its RHS, he will obtain the sum over all possible products over all possible prime numbers (each taken in the power z). Taking into account that any natural number can be obtained in the unique way by the product of its prime divisors one can see that this is nothing but the definition of the ζ-function.

3.3   Riemann's Hypothesis

Consider the integral

$$\int \frac{(-\xi)^{z-1}\, e^{-a\xi}}{1 - e^{-\xi}}\, d\xi,$$

over the contours C_N^* and C_N, which are defined on the figure:

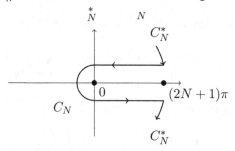

Note that $C = \lim_{N \to \infty} C_N$ is the contour that we have used in the integral representation above. In the region bounded by the joint contour C_N^* together with C_N the integrand under consideration is the analytic function with simple poles at $\xi = \pm 2\pi n i,\, 1 \le n \le N$. Then,

$$\frac{1}{2\pi i} \int_{C_N^*} \frac{(-\xi)^{z-1}\, e^{-a\xi}}{1 - e^{-\xi}}\, d\xi - \frac{1}{2\pi i} \int_{C_N} \frac{(-\xi)^{z-1}\, e^{-a\xi}}{1 - e^{-\xi}}\, d\xi = \sum_{n=1}^{N} \left(R_n + R_n' \right),$$

where R_n and R_n' are the residues of the integrand at $\xi = 2\pi n i$ and at $\xi = -2\pi n i$, correspondingly.

At the point where $\xi = 2\pi n\, e^{\pi i/2}$ the residue is equal to

$$(2\pi n)^{z-1}\, e^{-\frac{1}{2}\pi i\,(z-1)}\, e^{-2a\pi n i}.$$

Hence,

$$R_n + R_n' = (2\pi n)^{z-1}\, 2\,\sin\left(\frac{1}{2}\pi z + 2\pi n a \right)$$

and

$$-\frac{1}{2\pi i} \int_{C_N} \frac{(-\xi)^{z-1}\, e^{-a\xi}}{1 - e^{-\xi}}\, d\xi = \frac{2\,\sin\left(\frac{\pi z}{2} \right)}{(2\pi)^{1-z}} \sum_{n=1}^{N} \frac{\cos(2\pi n a)}{n^{1-z}} + \frac{2\,\cos\left(\frac{\pi z}{2} \right)}{(2\pi)^{1-z}} \sum_{n=1}^{N} \frac{\sin(2\pi n a)}{n^{1-z}}$$

$$- \frac{1}{2\pi i} \int_{C_N^*} \frac{(-\xi)^{z-1}\, e^{-a\xi}}{1 - e^{-\xi}}\, d\xi.$$

Then, because $0 < a \leq 1$ it is not hard to find N-independent number K such that

$$\left| e^{-a\xi} \left(1 - e^{-\xi}\right)^{-1} \right| < K,$$

when ξ is on C_N^*. As the result, we have that:

$$\left| \frac{1}{2\pi i} \int_{C_N^*} \frac{(-\xi)^{z-1} e^{-a\xi}}{1 - e^{-\xi}} \, d\xi \right| < \frac{K}{2\pi} \int_{-\pi}^{\pi} d\theta \, \left| (2N+1)^z \, \pi^z \, e^{iz\theta} \right|$$

$$< K \, (2N+1)^{\operatorname{Re} z} \, \pi^{\operatorname{Re} z} \, e^{\pi \operatorname{Re} z} \to 0,$$

as $N \to \infty$, if $\operatorname{Re} z < 0$.

Hence, for $\operatorname{Re} z < 0$ we obtain the so called Hurwitz formula:

$$\zeta(z, a) = \frac{2\,\Gamma(1-z)}{(2\pi)^{1-z}} \left[\sin\left(\frac{\pi z}{2}\right) \sum_{n=1}^{\infty} \frac{\cos(2\pi n a)}{n^{1-z}} + \cos\left(\frac{\pi z}{2}\right) \sum_{n=1}^{\infty} \frac{\sin(2\pi n a)}{n^{1-z}} \right].$$

Note that both series on the RHS of this expression are convergent.

If we put in this equation $a = 1$ and use the Eq. (2.2) and that $\sin(\pi z) = 2 \sin\left(\frac{\pi z}{2}\right) \cos\left(\frac{\pi z}{2}\right)$, we obtain the so called Riemann relation:

$$2^{1-z} \, \Gamma(z) \, \zeta(z) \, \cos\left(\frac{\pi z}{2}\right) = \pi^z \, \zeta(1-z),$$

which allows to formulate the Riemann's hypothesis.

We know that $\zeta(z)$ does not have zeros for $\operatorname{Re} z > 1$. From the Riemann's relation we see that zeros of $\zeta(z)$ in the region where $\operatorname{Re} z < 0$ will be those of the function

$$\frac{\sec\left(\frac{\pi z}{2}\right)}{\Gamma(z)},$$

i.e. the points $z = -2, -3, \ldots$. Thus, all other zeros of $\zeta(z)$ (except $z = -2, -3, -4, \ldots$) can only lay in the stripe $0 \leq \operatorname{Re} z \leq 1$ of the complex z-plane.

Riemann's hypothesis states that all zeros of $\zeta(z)$ in the stripe $0 \leq \operatorname{Re} z \leq 1$ are sitting on the line $\operatorname{Re} z = \frac{1}{2}$.

3.4 Application: Functional Determinant

Let us consider an application of the Riemann ζ-function. Suppose we would like to calculate the determinant of the one-dimensional Laplace operator d^2/dt^2 acting on the scalar functions that are defined on the interval $[0, T]$ and are vanishing at its ends:

$$I = \det \left[-\frac{d^2}{dt^2} \right]_{t \in [0, T]}.$$

First, let us specify what means such a determinant.

One can multiply functions by numbers and sum them up. This allows to define the vector space structures on the space of functions. The vector space can be infinite dimensional or even continual. It has a norm, which is defined by an integral that we specify in a moment. Vector space with a norm is referred to as the Hilbert space.

If the space of functions is a vector one, then differential operators can be understood as matrices in the same sense as a matrix operator acts on an ordinary finite dimensional vector space. In the present case the Hilbert space consists of such functions $f(t)$ on the interval $t \in [0, T]$ that are vanishing on its ends $f(0) = f(T) = 0$. Let us consider the eigen-functions of the differential operator in question:

$$-\frac{d^2 f(t)}{dt^2} = \lambda f(t) \quad \text{where} \quad f(0) = f(T) = 0.$$

It is not hard to see that the eigen-functions and eigen-values are as follows:

$$-\frac{d^2}{dt^2} \sin \left(\frac{\pi n t}{T} \right) = \left(\frac{\pi n}{T} \right)^2 \sin \left(\frac{\pi n t}{T} \right), \quad n \geq 1.$$

Obviously $\sin \left(\frac{\pi n t}{T} \right) = 0$ at $t = 0$ and $t = T$. At the same time, these functions compose the orthonormal,

$$\frac{2}{T} \int_0^T dt \, \sin \left(\frac{\pi n t}{T} \right) \sin \left(\frac{\pi m t}{T} \right) = \delta_{nm},$$

and complete,

$$\frac{2}{T} \sum_{n=1}^{+\infty} \sin \left(\frac{\pi n t}{T} \right) \sin \left(\frac{\pi n t'}{T} \right) = \delta \left(t - t' \right),$$

basis in the Hilbert space under consideration. Here $\sqrt{2/T}$ is the normalization coefficient of the mode functions. The first integral here defines the scalar product and, hence, the norm in the Hilbert space under consideration.

Thus, it is natural to define the determinant in question as the product of all the eigen-values of the operator ("matrix") under consideration:

$$\det \left[-\frac{d^2}{dt^2} \right]_{t \in [0, T]} \equiv \prod_{n=1}^{+\infty} \left(\frac{\pi n}{T} \right)^2.$$

Naively this is infinite, because

$$\prod_{n=1}^{+\infty}\left(\frac{\pi n}{T}\right)^2 = \left(\frac{\pi}{T}\right)^{2\sum_{n=1}^{+\infty}1} e^{2\sum_{n=1}^{+\infty}\log n}.$$

However, via analytical continuation one can define:

$$\zeta(0) = \sum_{n=1}^{+\infty}1, \quad \text{and} \quad \zeta'(0) = \sum_{n=1}^{+\infty}\log n.$$

Knowing the values of $\zeta(0) = -1/2$ and that $\zeta'(0)$ is a finite number, one can find that

$$\boxed{\det\left[-\frac{d^2}{dt^2}\right]_{t\in[0,T]} = \frac{T}{\pi}e^{2\zeta'(0)},}$$

which completes the calculation of the functional determinant in question.

There is one important point which is worth stressing here. Consider the function $\zeta(z, a)$ for the case when $a = 2$. Recall that $\zeta(z, 1) = \zeta(z)$. Then, on the one hand we have that

$$\zeta(z, 2) = \sum_{n=0}^{+\infty}\frac{1}{(n+2)^z} = \zeta(z) - 1.$$

On the other hand

$$\zeta(0, 2) = \sum_{n=0}^{+\infty}1.$$

Thus, taking into account previous discussion in this subsection, we have a puzzle whether we should take

$$\sum_{n=0}^{+\infty}1 = \zeta(0) \quad \text{or} \quad \sum_{n=0}^{+\infty}1 = \zeta(0) - 1 \ ?$$

The situation gets even worth if one considers

$$\zeta(z, 3) = \zeta(z) - 1 - \frac{1}{2^z}$$

or even $\zeta(z, n)$ for generic n. Note that while $\zeta(z) - \zeta(z, 2)$ is analytic in the entire complex z-plane (including infinity), the function $\zeta(z) - \zeta(z, n)$ is not analytic at infinity for $n \geq 3$.

So what should one do to resolve the puzzle? Following G. H. Hardy one implements axioms how to operate with divergent together with convergent series and defines that $\sum_{n=0}^{+\infty} 1 = \zeta(0)$. Then from this divergent series one obtains correct convergent ones. Other options for $\sum_{n=0}^{+\infty} 1$ lead to contradictory consequences.

Chapter 4
Hermite Polynomials

Abstract This section is recorded by MIPT student Sharipov Rustem. It contains the derivation of the properties of the Hermite polynomials and their application to quantum mechanics and representation theory.

4.1 Application: Schrödinger Equation

The Hamiltonian for the classical oscillator with the unit frequency and mass is as follows:

$$\mathcal{H} = \frac{1}{2} (p^2 + q^2),$$

where the Poisson brackets for p and q are defined as

$$\{q, \, p\} = 1.$$

In passing to quantum oscillator one changes p and q for the corresponding operators $\hat{p}$ and $\hat{q}$, which, when acting on the functions of x, have the following representation $\hat{q} \rightarrow x$ and $\hat{p} \rightarrow -i\,\hbar\,d/dx$. Then, the Poisson brackets are transformed into the commutation relations

$$\left[\hat{q}, \, \hat{p}\right] \equiv \hat{q}\,\hat{p} - \hat{p}\,\hat{q} = i\,\hbar,$$

and $\mathcal{H}$ is transformed into the quantum oscillator Hamiltonian:

$$\hat{\mathcal{H}} = \frac{1}{2} \left(\hat{p}^2 + \hat{q}^2\right).$$

From now on we set the Planck constant $\hbar$ to one.

© The Author(s), under exclusive license to Springer Nature Switzerland AG 2019
V. Akhmedova and E. T. Akhmedov, *Selected Special Functions
for Fundamental Physics*, SpringerBriefs in Physics,
https://doi.org/10.1007/978-3-030-35089-5_4

As the result one defines the quantum oscillator Schrödinger equation as:

$$\hat{\mathcal{H}} \psi \equiv \frac{1}{2} \left[-\frac{d^2}{dx^2} + x^2 \right] \psi = E \psi,$$

where E is the energy or an eigen-value of the Hamiltonian operator.

The Hilbert space, on which the Hamiltonian operator acts, consists of normalizable functions $\psi(x)$ on the real line $x \in \mathbb{R}$,

$$\int\limits_{-\infty}^{+\infty} dx \, |\psi(x)|^2 < \infty,$$

which in general are taking complex values; $\psi(x)$ is the so called quantum mechanical wave function and by definition $|\psi(x)|^2 \, dx$ is the probability density to find the quantum particle in the interval $[x, \, x + dx]$. Hence, if it is properly normalized, the above integral should be equal to one.

The Hamiltonian operator under consideration is Hermitian, $\hat{\mathcal{H}}^+ = \hat{\mathcal{H}}$, when acts on such functions, which by definition means that:

$$\int\limits_{-\infty}^{+\infty} dx \, \psi_1^*(x) \left[\hat{\mathcal{H}} \psi_2(x) \right] \equiv \int\limits_{-\infty}^{+\infty} dx \, \left[\hat{\mathcal{H}}^+ \, \psi_1^*(x) \right] \psi_2(x) = \int\limits_{-\infty}^{+\infty} dx \, \left[\hat{\mathcal{H}} \psi_1^*(x) \right] \psi_2(x),$$

for arbitrary two functions $\psi_{1,2}$ from the Hilbert space under consideration.

Let us define now the following operators

$$\hat{a} = \frac{1}{\sqrt{2}} \left[\frac{d}{dx} + x \right] = \frac{1}{\sqrt{2}} \left[i \, \hat{p} + \hat{q} \right] \quad \text{and} \quad \hat{a}^+ = \frac{1}{\sqrt{2}} \left[-\frac{d}{dx} + x \right] = \frac{1}{\sqrt{2}} \left[-i \, \hat{p} + \hat{q} \right].$$

(4.1)

They obey the so called Heisenberg algebra:

$$\left[\hat{a}, \, \hat{a}^+ \right] \equiv \hat{a} \, \hat{a}^+ - \hat{a}^+ \, \hat{a} = 1,$$

(4.2)

which can be shown by the direct calculation. It is related to the above defined commutation relations for $\hat{p}$ and $\hat{q}$.

Furthermore, it is straightforward to see that:

$$\hat{a} \, \hat{a}^+ - \frac{1}{2} = \hat{a}^+ \hat{a} + \frac{1}{2} = \frac{1}{2} \left[-\frac{d^2}{dx^2} + x^2 \right] \equiv \hat{\mathcal{H}}.$$

Thus, the Schrödinger equation under consideration can be rewritten as:

$$\left[\hat{a}^+\hat{a} + \frac{1}{2}\right]\psi = E\,\psi. \tag{4.3}$$

To find its solutions consider the equation $\hat{a}\psi_0 = 0$. We want to find such a ψ_0, which obeys the normalization condition:

$$\int\limits_{-\infty}^{+\infty} dx\, |\psi_0(x)|^2 = 1,$$

because, as we have mentioned above, the total probability to find the particle some-where should be equal to one.

From the equation under consideration,

$$\hat{a}\psi_0 = \frac{1}{\sqrt{2}}\left[\frac{d}{dx} + x\right]\psi_0 = 0,$$

we obtain that

$$\psi_0 = N_0 e^{-\frac{x^2}{2}},$$

where N_0 is the normalization constant. From the above normalization condition one can find this constant:

$$\psi_0(x) = \frac{1}{\pi^{\frac{1}{4}}} e^{-\frac{x^2}{2}}.$$

Thus, from the above considerations it follows that:

$$\left[\hat{a}^+\hat{a} + \frac{1}{2}\right]\psi_0 = \frac{1}{2}\,\psi_0,$$

which means that we have found here a solution of the Schrödinger equation (4.3) corresponding to the Hamiltonian eigen-value (or energy) equal to $E_0 = 1/2$.

Consider now the function

$$\psi_1(x) = N_1\hat{a}^+\psi_0(x) = N_1\frac{\sqrt{2}x}{\pi^{\frac{1}{4}}}e^{-\frac{x^2}{2}},$$

where the constant N_1 again follows from the normalization condition:

$$\int\limits_{-\infty}^{+\infty} dx\, |\psi_1(x)|^2 = 1.$$

It is straightforward to see that:

$$\left[\hat{a}^+\hat{a} + \frac{1}{2}\right]\psi_1 = \left[\hat{a}^+\hat{a} + \frac{1}{2}\right]\hat{a}^+\psi_0 = \hat{a}^+\hat{a}\hat{a}^+\psi_0 + \frac{1}{2}\hat{a}^+\psi_0 =$$

$$= \hat{a}^+\psi_0 + (\hat{a}^+)^2\hat{a}\psi_0 + \frac{1}{2}\hat{a}^+\psi_0 = \frac{3}{2}\hat{a}^+\psi_0 = \frac{3}{2}\psi_1,$$

where we have used the commutation relations (4.2) and the equation for $\psi_0(x)$:
$\hat{a}\,\psi_0 = 0$.

Hence, $\psi_1(x)$ is another solution of the Schrödinger equation with a different
energy, which is equal to $E_1 = 3/2$. Similarly one can define:

$$\boxed{\psi_n(x) \equiv N_n\,(\hat{a}^+)^n e^{-\frac{x^2}{2}}, \quad n \in \mathbb{N},}$$

where N_n again can be found from the normalization condition. By induction it is
not hard to show that

$$\left[\hat{a}^+\hat{a} + \frac{1}{2}\right]\psi_n = \left(n + \frac{1}{2}\right)\psi_n,$$

i.e. $\psi_n(x)$ solves Eq. (4.3) with $E = E_n \equiv \left(n + \frac{1}{2}\right)$.

We will show below that

$$\psi_n(x) \equiv \frac{N_n}{2^{n/2}}\left[-\frac{d}{dx} + x\right]^n e^{-\frac{x^2}{2}} = N'_n\,H_n(x)\,e^{-\frac{x^2}{2}},$$

where $H_n(x)$ are the so called Hermite polynomials. Also we will explicitly find the
normalization constants N'_n below.

4.2 Definition

Consider again the chain of equations:

$$\frac{1}{2}\left[-\frac{d^2}{dx^2} + x^2\right]\psi_n(x) = \left(n + \frac{1}{2}\right)\psi_n(x), \quad n \in \mathbb{N}.$$

Represent the solutions of these equations as $\psi_n(x) = N'_n\,e^{-\frac{x^2}{2}}\,H_n(x)$, where N'_n are
some constants. Then substituting this ansatz into the last equation, one can find that
H_n obeys the following equation:

$$\boxed{H''_n - 2x\,H'_n + 2n\,H_n = 0.} \tag{4.4}$$

As we will show below, solutions of this chain of equations are the Hermite polynomials, which can be represented as:

$$H_n(x) = (-1)^n e^{x^2} \frac{d^n}{dx^n} e^{-x^2}.$$ (4.5)

Explicitly some of the polynomials for small values of n are as follows

$$H_0(x) = 1, \quad H_1(x) = 2x, \quad H_2(x) = 4x^2 - 2 \quad \text{and} \quad H_3(x) = 8x^3 - 12x, \ ...$$

which can be deduced from (4.5) by the direct calculation.

4.3 Generating Function

To show that the function defined by (4.5) indeed solves (4.4) consider the following function:

$$W(x, t) = e^{2xt - t^2} = \sum_{n=0}^{+\infty} \frac{c_n(x)}{n!} t^n, \quad |t| < +\infty.$$

Here

$$c_n(x) = \frac{\partial^n W}{\partial t^n}\bigg|_{t=0}.$$

But

$$\frac{\partial^n W}{\partial t^n}\bigg|_{t=0} = e^{x^2} \left[\frac{\partial^n}{\partial t^n} e^{-(x-t)^2} \right]_{t=0} = (-1)^n e^{x^2} \frac{d^n}{du^n} e^{-u^2}\bigg|_{u=x} \equiv H_n(x).$$

Thus, $W(x, t)$ is the so called generating function of the Hermite polynomials:

$$e^{2xt - t^2} = \sum_{n=0}^{\infty} \frac{H_n(x)}{n!} t^n, \quad |t| < \infty.$$ (4.6)

Putting $x = 0$ in this equation, one can find that

$$H_{2n}(0) = (-1)^n \frac{(2n)!}{n!} \quad \text{and} \quad H_{2n+1}(0) = 0,$$

which is useful in applications.

4.4   Recurrence Relations

Generating function is convenient for the derivation of the so called recurrence relations obeyed by the polynomials, which, in their own right, allow to derive the
differential equations for the polynomials, as we will see in a moment.

The generating function obeys the following relation:

$$\frac{\partial W}{\partial t} - (2x - 2t)\, W = 0,$$

which can be found by the direct calculation, using the explicit form of $W(x, t)$
provided above. On the other hand, substitution of the expansion (4.6) into this
equation leads to:

$$\sum_{n=0}^{\infty} \frac{H_{n+1}(x)}{n!} t^n - 2x \sum_{n=0}^{\infty} \frac{H_n(x)}{n!} t^n + 2 \sum_{n=0}^{\infty} \frac{H_n(x)}{n!} t^{n+1} = 0.$$

Then, equating to zero the coefficients multiplying t^n for each n separately, we find
the first recurrence relation:

$$\boxed{H_{n+1}(x) - 2x\, H_n(x) + 2n\, H_{n-1}(x) = 0. \quad n = 1, 2, \dots .}$$ (4.7)

Another equation that the function $W(x, t)$ does obey can be also found by the direct
calculation using its explicit form:

$$\frac{\partial W}{\partial x} - 2t\, W = 0.$$

Then, substituting into it the expansion (4.6), we obtain:

$$\sum_{n=0}^{\infty} \frac{H_n'(x)}{n!} t^n - 2 \sum_{n=0}^{\infty} \frac{H_n(x)}{n!} t^{n+1} = 0,$$

where the prime means the differentiation with respect to x.

Also equating to zero the coefficients multiplying t^n for each n separately, we find
another recurrence relation obeyed by the polynomials under consideration:

$$\boxed{H_n'(x) = 2n\, H_{n-1}(x), \quad n = 1, 2, \dots .}$$

Excluding $H_{n-1}(x)$ from the both recurrence relations obtained above, we find the
new one:

$$\boxed{H_{n+1}(x) - 2x\, H_n(x) + H_n'(x) = 0, \quad n \in \mathbb{N}.}$$

Differentiating this equation by x and using again the above two recurrence relations, we find that the function $H_n(x)$ does obey the equation for the Hermite polynomials (4.4). In all, we have proven that the functions $\psi_n(x)$ expressed via the polynomials $H_n(x)$ indeed obey the Schrödinger equation for the quantum oscillator.

4.5 Integral Representation

Below it will be convenient to use the integral representation for the Hermite polynomials. To find it let us use the following relation:

$$e^{-x^2} = \frac{1}{\sqrt{\pi}} \int\limits_{-\infty}^{+\infty} e^{-t^2} e^{i\,2xt}\, dt,$$

which can be proved by the direct calculation of the standard Gaussian integral.

Differentiating this equation n times, we straightforwardly obtain that:

$$H_n(x) = \frac{2^n(-i)^n e^{x^2}}{\sqrt{\pi}} \int\limits_{-\infty}^{+\infty} dt\, t^n e^{-t^2+2ixt}, \quad n \in \mathbb{N}. \tag{4.8}$$

This is the integral representation, which we will use below.

4.6 Fourier Transformation

Let us change in the function $W(x,t)$ its argument x into p and multiply the obtained function $W(p,t)$ by $e^{ipx-\frac{p^2}{2}}$. Integrating the resulting expression over $p \in (-\infty, +\infty)$, one obtains:

$$\int\limits_{-\infty}^{+\infty} e^{2pt-t^2-\frac{p^2}{2}+ipx}\, dp = \int\limits_{-\infty}^{+\infty} dp\, e^{ipx-\frac{p^2}{2}} \sum_{n=0}^{\infty} \frac{H_n(p)}{n!} t^n = \sum_{n=0}^{\infty} \frac{t^n}{n!} \int\limits_{-\infty}^{+\infty} e^{ipx-\frac{p^2}{2}} H_n(p)\, dp.$$

$$\tag{4.9}$$

Explicitly calculating the integral on the LHS of this relation, we find that:

$$\int\limits_{-\infty}^{+\infty} e^{2pt-t^2-\frac{p^2}{2}+ipx}\, dp = \sqrt{2\pi}\, e^{t^2+2ixt-\frac{x^2}{2}} = \sqrt{2\pi}\, e^{-\frac{x^2}{2}} \sum_{n=0}^{\infty} \frac{(it)^n}{n!} H_n(x).$$

Then, equating the coefficients multiplying t^n on both sides of (4.9) for each n separately, we obtain the relation:

$$e^{-\frac{x^2}{2}} H_n(x) = \frac{1}{i^n \sqrt{2\pi}} \int\limits_{-\infty}^{+\infty} dp \, e^{ipx} e^{-\frac{p^2}{2}} H_n(p), \quad n \in \mathbb{N},$$

which can be written as:

$$\psi_n(x) = \frac{1}{i^n \sqrt{2\pi}} \int\limits_{-\infty}^{+\infty} dp \, e^{i\,px} \, \psi_n(p),$$

where $\psi_n(x)$ is the solution of the quantum oscillator Schrödinger equation.

The obtained equation shows that the Fourier transformation of the function $\psi_n(x)$ is equal to the same function, $\psi_n(p)$, with the exchanged argument, $x \to p$. That is because the oscillator Hamiltonian with the unit frequency and mass $\hat{\mathcal{H}} = \frac{1}{2} \left(\hat{p}^2 + \hat{q}^2 \right)$ is invariant under the exchange of $\hat{p}$ and $\hat{q}$ with each other. Furthermore, under the Fourier transformation the Hamiltonian changes as: $\frac{1}{2} \left[-\frac{d^2}{dx^2} + x^2 \right] \to \frac{1}{2} \left[p^2 - \frac{d^2}{dp^2} \right]$, i.e. the obtained here relation can be foreseen already from the initial equation under consideration.

4.7   Orthogonality

As we have mentioned above the Eq. (4.3) can be understood as the one defining eigen-functions of the quantum oscillator Hamiltonian. The latter is the Hermitian operator and its eigen-functions compose the complete and orthonormal basis in the corresponding Hilbert space. The situation is pretty much like the one with non-degenerate matrices and vector spaces. Our goal in the next few sections is to see these facts explicitly.

Rewrite the Schrödinger differential equation for $\psi_n(x) = N'_n e^{-\frac{x^2}{2}} H_n(x)$ as:

$$\psi_n'' + \left(2n + 1 - x^2 \right) \psi_n = 0.$$

Multiply this equation by $\psi_m(x)$. Then subtract from the obtained expression the equation

$$\psi_m'' + \left(2m + 1 - x^2 \right) \psi_m = 0,$$

multiplied by $\psi_n(x)$. This way we obtain the following relation:

$$\frac{d}{dx}\left(\psi_n'\psi_m - \psi_m'\psi_n\right) + 2\,(n-m)\,\psi_n\psi_m = 0.$$

Integrating this relation over $x \in (-\infty, +\infty)$, one finds that:

$$(n-m)\int_{-\infty}^{+\infty}\psi_n(x)\,\psi_m(x)\,dx = 0.$$

Thus, we obtain the following orthogonality relation for the Hermite polynomials:

$$\boxed{\int_{-\infty}^{+\infty} dx\, e^{-x^2}\, H_n(x)\, H_m(x) = 0, \quad \text{if}\ \ n \neq m.} \qquad (4.10)$$

To find the normalization coefficients N_n', which have been defined above, change in Eq. (4.7) n for $n-1$ and multiply the resulting expression by $H_n(x)$. Then subtract from the obtained equation the Eq. (4.7) itself multiplied by $H_{n-1}(x)$. The result of these manipulations is as follows:

$$H_n^2(x) + 2\,(n-1)\,H_n(x)\,H_{n-2}(x) - H_{n+1}(x)\,H_{n-1}(x) - 2\,n\,H_{n-1}^2(x) = 0,$$

where $n = 2, 3, \ldots$

Multiplying this relation by e^{-x^2} and integrating it over $x \in (-\infty, +\infty)$, one can find that:

$$\int_{-\infty}^{+\infty} dx\, e^{-x^2}\, H_n^2(x) = 2n \int_{-\infty}^{+\infty} e^{-x^2}\, H_{n-1}^2(x)\, dx, \quad \text{where}\ \ n = 1, 2, 3, \ldots .$$

To obtain this relation we have used Eq. (4.10).

Applying the last relation several times to reduce n on the RHS, we find that:

$$\int_{-\infty}^{+\infty} e^{-x^2}\, H_n^2(x)\, dx = 2^n\, n! \int_{-\infty}^{+\infty} e^{-x^2}\, H_0^2(x)\, dx = 2^n\, n!\, \sqrt{\pi}, \quad n \in \mathbb{N},$$

where on the last step we have used explicitly that $H_0(x) = 1$ and took the Gaussian integral.

Hence, if we define $\psi_n(x)$ functions as

$$\psi_n(x) = \frac{H_n(x)\,e^{-\frac{x^2}{2}}}{\left(2^n\,n!\,\sqrt{\pi}\right)^{\frac{1}{2}}},$$

then they compose the orthonormal basis of functions on the real line $x \in (-\infty, +\infty)$. The last equation obviously defines the normalization constants N'_n, which we have introduced above.

4.8   Asymptotic Form for the Large Index

To show the completeness of the obtained basis of functions we need to find the asymptotic form of the Hermite polynomials as $n \to \infty$. After an obvious change of integration variables in the integral representation (4.8) one can represent the Hermite polynomials as:

$$H_n(x) = \frac{2^n}{\sqrt{\pi}} \int\limits_{-\infty}^{+\infty} dt\,(x - it)^n e^{-t^2}.$$

To estimate this integral as $n \to \infty$ we will use the steepest descent or stationary phase approximation method. We have used it already in the section on Γ-functions to derive the Stirling formula. In the present concrete case we have to solve the equation

$$\left[-t^2 + n\log(x - it)\right]' = 0,$$

to find the extrema. Hence,

$$2t + \frac{ni}{x - it} = 0,$$

and we have two extrema. At the same time:

$$\left[-t^2 + n\log(x + it)\right]'' = -2 + \frac{n}{(x - it)^2}.$$

Then, using both extrema, we obtain that:

$$H_n(x) \approx \sqrt{2}\left(\frac{2n}{e}\right)^{\frac{n}{2}} e^{\frac{x^2}{2}} \cos\left(\sqrt{2n+1}\,x - \frac{\pi n}{2}\right), \quad \text{as } n \to \infty. \qquad (4.11)$$

This expression defines the asymptotic form of the Hermite polynomials for large values of their index.

4.9 Completeness

Now we are ready to show the completeness of the basis of solutions of the quantum oscillator equation. Consider the recurrence relation (4.7), multiply it by $H_n(y)$ and, then, subtract from the obtained expression the same one with x and y exchanged with each other. As the result we obtain:

$$\left[H_{n+1}(x) H_n(y) - H_{n+1}(y) H_n(x)\right] - 2n\left[H_n(x) H_{n-1}(y) - H_n(y) H_{n-1}(x)\right] =$$

$$= 2(x - y) H_n(x) H_n(y), \quad n = 1, 2, \ldots .$$

Dividing this expression by $2^n n!$, then summing it over n from 1 to m, and using that $H_0(x) = 1$ and $H_1(x) = 2x$, one can find the following relation:

$$2(x - y) \sum_{n=1}^{m} \frac{H_n(x) H_n(y)}{2^n n!} = \frac{H_{m+1}(x) H_m(y) - H_{m+1}(y) H_m(x)}{2^m m!} - 2(x - y).$$

It can be rewritten as:

$$\boxed{\frac{1}{\sqrt{\pi}} \sum_{n=0}^{m} \frac{H_n(x) H_m(y)}{2^n n!} = \frac{H_{m+1}(x) H_m(y) - H_{m+1}(y) H_m(x)}{(x - y) 2^{m+1} m! \sqrt{\pi}}.}$$

Now taking the limit $m \to \infty$ and using on the RHS of this expression the asymptotic form (4.11), one can show that

$$\frac{1}{\sqrt{\pi}} \sum_{n=0}^{\infty} \frac{e^{-\frac{x^2}{2}} H_n(x) e^{-\frac{y^2}{2}} H_n(y)}{2^n n!} = \delta(x - y). \tag{4.12}$$

To obtain the last equation we have used Eq. (4.11) and that as $m \to \infty$:

$$\cos\left[\sqrt{2(m+1)}\, x - \frac{\pi(m+1)}{2}\right] \cos\left[\sqrt{2m}\, y - \frac{\pi m}{2}\right] -$$

$$- \cos\left[\sqrt{2(m+1)}\, y - \frac{\pi(m+1)}{2}\right] \cos\left[\sqrt{2m}\, x - \frac{\pi m}{2}\right] =$$

$$= \sin\left[\sqrt{2(m+1)}\, x - \frac{\pi m}{2}\right] \cos\left[\sqrt{2m}\, y - \frac{\pi m}{2}\right] -$$

$$- \sin\left[\sqrt{2(m+1)}\, y - \frac{\pi m}{2}\right] \cos\left[\sqrt{2m}\, x - \frac{\pi m}{2}\right] \approx$$

$$\approx \sin\left[\sqrt{2m}\,x - \frac{\pi(m+1)}{2}\right]\cos\left[\sqrt{2m}\,y - \frac{\pi m}{2}\right] -$$

$$-\sin\left[\sqrt{2m}\,y - \frac{\pi(m+1)}{2}\right]\cos\left[\sqrt{2m}\,x - \frac{\pi m}{2}\right] = \sin\left[\sqrt{2m}\,(x-y)\right].$$

To obtain the δ-function on the RHS of (4.12) we have used the following resolution of the δ-function:

$$\delta(x-y) = \lim_{T\to\infty}\int_{-T}^{T} e^{ip(x-y)}dp = \lim_{T\to\infty}\frac{e^{ip(x-y)}}{i(x-y)}\bigg|_{-T}^{T} = \lim_{T\to\infty}\frac{2\sin[T(x-y)]}{(x-y)}.$$

In other terms the relation (4.12) states that

$$\boxed{\sum_{n=0}^{\infty}\psi_n(x)\,\psi_n(y) = \delta(x-y),}\qquad(4.13)$$

which is the condition of the completeness of the basis of functions $\psi_n(x)$.

To understand our statements about orthonormality and completeness in the space of functions, recall that if one has a complete and orthonormal basis of vectors in a D-dimensional space, i.e. $\vec{e}_a = e_a^i$, $a = \overline{1, D}$ and $i = \overline{1, D}$, then the condition of its orthonormality states that

$$(\vec{e}_a,\,\vec{e}_b) = \delta_{ab}.$$

This relation is the finite dimensional analog of (4.10). At the same time the condition of the completeness of such a basis states that:

$$\sum_{a=1}^{D} e_a^i\, e_a^j = \delta^{ij}.$$

The last relation, in its own right is the finite dimensional analog of (4.12) and (4.13). In fact, multiplying its both sides by v^i we obtain the decomposition of the vector $\vec{v}$:

$$\vec{v} = \sum_{a=1}^{D} v_a\,\vec{e}_a,$$

where $v_a = (\vec{v},\,\vec{e}_a)$ is the projection of the vector $\vec{v}$ on the basis element $\vec{e}_a$.

4.10   Relation to the Representations of the Heisenberg Algebra

Frequently a chain of functions solving a chain of differential equations (such as e.g. (4.3)) composes a representation of a symmetry algebra present in the problem (sometimes inherently). We will encounter such situations in many cases below.

To see this phenomenon, consider one of the simplest examples. Namely, consider the differential equation for the classical oscillator with the unit frequency:

$$\frac{d^2 u(\varphi)}{d\varphi^2} = -u(\varphi).$$

The basis of solutions of this equation consists of $u_1 = \sin \varphi$ and $u_2 = \cos \varphi$. These functions obviously obey the following relations:

$$\frac{d}{d\varphi} \cos \varphi = -\sin \varphi, \quad \text{and} \quad \frac{d}{d\varphi} \sin \varphi = \cos \varphi.$$

These relations can be written as:

$$\frac{d}{d\varphi} \begin{pmatrix} \cos \varphi \\ \sin \varphi \end{pmatrix} = \begin{pmatrix} 0 & -1 \\ 1 & 0 \end{pmatrix} \begin{pmatrix} \cos \varphi \\ \sin \varphi \end{pmatrix}. \tag{4.14}$$

The differential operator $d/d\varphi$, for $\varphi \in [0, 2\pi)$, is the so called generator in the algebra of rotations of the two dimensional plane, i.e. of the $SO(2)$ algebra. We explain this in a moment.

In these notes we prefer to give explicit constructive examples rather than formal definitions. But to move further let us define here what is Lie algebra. An example of such an algebra is the $SO(2)$. An algebra is a vector space, elements of which can be multiplied by numbers, added and multiplied to each other. Every algebra contains a null element, which, if added to any other element of algebra, does not change it.

Lie algebra is such an algebra which instead of the product of its elements is equipped with the so called Lie brackets. Namely if a and b are elements of a Lie algebra g, then their Lie brackets $[a, b]$ also belong to the same algebra. There is a unit or null element in the algebra which gives vanishing Lie brackets with any element of the algebra. The Lie brackets by definition should obey the following properties:

- Anti-symmetry: $[a, b] = -[b, c]$;
- The Jacoby identity:

$$[a, [b, c]] + [b, [c, a]] + [c, [a, b]] = 0.$$

An explicit example of the Lie brackets is the commutator.

Now let us return to the concrete situation under consideration. Consider the two-dimensional unit vector $(\cos \varphi, \sin \varphi)$. Under the rotation of the plane by an angle ϕ it changes as:

$$\begin{pmatrix} \cos(\varphi + \phi) \\ \cos(\varphi + \phi) \end{pmatrix} = \begin{pmatrix} \cos \phi & -\sin \phi \\ \sin \phi & \cos \phi \end{pmatrix} \begin{pmatrix} \cos \varphi \\ \sin \varphi \end{pmatrix}.$$

For infinitesimal angle, $\phi \ll 1$, we have that:

$$\begin{pmatrix} \cos(\varphi + \phi) \\ \sin(\varphi + \phi) \end{pmatrix} \approx \begin{pmatrix} \cos \varphi \\ \sin \varphi \end{pmatrix} + \phi \begin{pmatrix} -\sin \varphi \\ \cos \varphi \end{pmatrix} =$$
$$= \left[\begin{pmatrix} 1 & 0 \\ 0 & 1 \end{pmatrix} + \phi \begin{pmatrix} 0 & -1 \\ 1 & 0 \end{pmatrix} \right] \begin{pmatrix} \cos \varphi \\ \sin \varphi \end{pmatrix} = \left[1 + \phi \frac{d}{d\varphi} \right] \begin{pmatrix} \cos \varphi \\ \sin \varphi \end{pmatrix}, \quad (4.15)$$

where to write the last equality we have used the relation (4.14).

Finally, one can show the following chain of relations:

$$\begin{pmatrix} \cos \phi & -\sin \phi \\ \sin \phi & \cos \phi \end{pmatrix} = \cos \phi \begin{pmatrix} 1 & 0 \\ 0 & 1 \end{pmatrix} + \sin \phi \begin{pmatrix} 0 & -1 \\ 1 & 0 \end{pmatrix} = \exp \left\{ \begin{pmatrix} 0 & -1 \\ 1 & 0 \end{pmatrix} \phi \right\} =$$
$$= \begin{pmatrix} 1 & 0 \\ 0 & 1 \end{pmatrix} + \phi \begin{pmatrix} 0 & -1 \\ 1 & 0 \end{pmatrix} + \ldots \quad \Leftrightarrow \quad e^{\phi \frac{d}{d\varphi}} = 1 + \phi \frac{d}{d\varphi} + \ldots \quad \Leftrightarrow \quad e^{i\phi} = 1 + i\phi + \ldots .$$

To understand this chain of relations recall that the exponent of a matrix $\hat{M}$ is just the following Taylor series:

$$e^{\hat{M}} = \hat{1} + \hat{M} + \frac{\hat{M}^2}{2} + \cdots + \frac{\hat{M}^n}{n!} + \ldots ,$$

and note that:

$$\begin{pmatrix} 0 & -1 \\ 1 & 0 \end{pmatrix}^2 = - \begin{pmatrix} 1 & 0 \\ 0 & 1 \end{pmatrix},$$

i.e. the matrix under consideration acts as the imaginary unit $i^2 = -1$. Then use the relation (4.14). And finally, the group of rotations acts by the multiplication of the phase $e^{i\phi}$, if one represents the two-dimensional unit vector $(\cos \varphi, \sin \varphi)$ as the complex number of the following form $\cos \varphi + i \sin \varphi = e^{i \varphi}$: $e^{i\phi} e^{i \varphi} = e^{i (\varphi + \phi)}$.

In all, the chain of functions $\cos \varphi$ and $\sin \varphi$, which represents the two-dimensional basis of solutions of the classical oscillator equation, provides a representation of the $SO(2)$ algebra. And the generator of this algebra can be represented either as imaginary unit i, if the algebra acts on $e^{i\varphi}$, or as the antisymmetric matrix in Eq. (4.14) or as the differential operator $d/d\varphi$, if the algebra acts on the unit vector.[1]

---

[1] Similarly $(\cosh t, \sinh t)$ solves the equation $\frac{d^2 u(t)}{dt^2} = u(t)$ and provides the representation of the $SO(1, 1)$ algebra—the algebra of Lorentz transformations in the two-dimensional Minkowski space-time.

Similar but a bit more complicated situation appears in the case of Hermite polynomials. Namely, we have the chain of the differential equations (4.3) for various $E_n = n + 1/2$, $n \in \mathbb{N}$ and the Heisenberg algebra (4.2). The operators $\hat{a}$ and $\hat{a}^+$ are the generators of the Lie algebra, i.e. are analogs of the operator that generates rotations. In the previous case on top of the unit or null operator we had only one generator of the algebra—the operator which provides the rotation by an infinitesimal angle: either $d/d\varphi$ or the

$$\begin{pmatrix} 0 & -1 \\ 1 & 0 \end{pmatrix}$$

matrix, depending on the representation of the algebra. Hence, in the previous case to define the algebra we did not provide any commutation relations, because there were no any non-trivial ones. In the present case, however, we have two operators $\hat{a}$ and $\hat{a}^+$, on top of the unit one, which do not commute with each other.

In all, in the present case we have the vector

$$[\psi_0(x), \psi_1(x), \psi_2(x), \ldots]$$

in the infinite dimensional Hilbert space, which provides the representation of the Heisenberg algebra (4.2). In fact, from the discussion of the Sects. 4.1 and 4.7 one can deduce that

$$\hat{a}\,\psi_n(x) = \sqrt{n}\,\psi_{n-1}(x) \quad \text{and} \quad \hat{a}^+\psi_n(x) = \sqrt{n+1}\,\psi_{n+1}(x).$$

Which means that the $\hat{a}$ and $\hat{a}^+$ operators can be represented as the following half-infinite matrices:

$$\hat{a}\begin{pmatrix} \psi_0 \\ \psi_1 \\ \psi_2 \\ \cdots \end{pmatrix} = \begin{pmatrix} 0 & 0 & 0 & \ldots \\ \sqrt{1} & 0 & 0 & \ldots \\ 0 & \sqrt{2} & 0 & \ldots \\ \cdot & \cdot & \cdot & \ldots \end{pmatrix} \begin{pmatrix} \psi_0 \\ \psi_1 \\ \psi_2 \\ \cdots \end{pmatrix}, \quad \hat{a}^+\begin{pmatrix} \psi_0 \\ \psi_1 \\ \psi_2 \\ \cdots \end{pmatrix} = \begin{pmatrix} 0 & \sqrt{1} & 0 & \ldots \\ 0 & 0 & \sqrt{2} & \ldots \\ 0 & 0 & 0 & \ldots \\ \cdot & \cdot & \cdot & \ldots \end{pmatrix} \begin{pmatrix} \psi_0 \\ \psi_1 \\ \psi_2 \\ \cdots \end{pmatrix}$$

when they act on the infinite dimensional vector under consideration. At the same time, it is not hard to see that the same algebra can be realized via either the operators $\hat{a} \to d/d\alpha$ and $\hat{a}^+ \to \alpha$: $[d/d\alpha, \alpha] = 1$ or via the operators (4.1), when it acts on the functions of α or x, correspondingly. The latter are other representations of the same algebra formally defined by the commutation relations of abstract operators $\hat{a}$ and $\hat{a}^+$ (4.2).

4.11   Applications: Back to the Quantum Oscillator

Most of the equations of this section can be reformulated in terms of the so called Dirac's bra and ket vectors of an abstract Hilbert space. Namely instead of functions one can introduce ket vectors, $|\psi\rangle$, while instead of the conjugate functions one can introduce the bra vectors $\langle\psi|$. E.g. the eigen-vectors of the Hamiltonian operator under consideration are defined as follows:

$$\hat{\mathcal{H}}\,|n\rangle = \left(n + \frac{1}{2}\right)\,|n\rangle.$$

At the same time one can introduce eigen-vectors of the $\hat{p}$ and $\hat{q}$ operators:

$$\hat{p}\,|k\rangle = k\,|k\rangle \quad\text{and}\quad \hat{q}\,|x\rangle = x\,|x\rangle.$$

Then, their scalar products by definition are:

$$\langle k|k'\rangle = \delta\left(k - k'\right), \quad\text{and}\quad \langle x|x'\rangle = \delta\left(x - x'\right), \tag{4.16}$$

which just means that the eigen-vectors $|k\rangle$ and $|x\rangle$ of the Hermitian operators $\hat{p}$ and $\hat{q}$, correspondingly, compose an orthogonal basis. Their completeness will be discussed below.

Furthermore:

$$k\,\langle x|k\rangle = \langle x|\hat{p}|k\rangle = -i\,\frac{d}{dx}\,\langle x|k\rangle.$$

While to obtain the left hand side of this relation we have used that $\hat{p}\,|k\rangle = k\,|k\rangle$, to find the right hand side of this relation we have used that $\langle x|\,\hat{p} = -i\,\frac{d}{dx}\,\langle x|$.

Hence, solving found this way differential equation for $\langle x|k\rangle$, one gets that:

$$\langle x|k\rangle \propto e^{ikx} \quad\text{and}\quad \langle k|x\rangle \propto e^{-ikx},$$

where the coefficients of proportionality here follow from the proper normalization set by Eq. (4.16). In fact, the condition of the completeness of the basis of vectors $|x\rangle$ for all x looks as:

$$\int\limits_{-\infty}^{+\infty} dx\,|x\rangle\langle x| = \hat{\mathbf{1}},$$

where $\hat{\mathbf{1}}$ is the unit operator. Similar condition is also true for the complete basis of vectors $|k\rangle$.

If one multiplies the last relation by $\langle k|$ from the left and by $|k'\rangle$ from the right, he well obtain that

$$\int\limits_{-\infty}^{+\infty} dx\,\langle k|x\rangle\langle x|k'\rangle = \langle k|\hat{\mathbf{1}}|k'\rangle = \langle k|k'\rangle = \delta(k - k').$$

But we know that

$$\int\limits_{-\infty}^{+\infty} \frac{dx}{2\pi}\, e^{-i\,(k-k')x} = \delta(k - k').$$

Hence, one can establish that e.g.

$$\langle x|k \rangle = \frac{e^{ikx}}{\sqrt{2\pi}},$$

which fixes the coefficient in the relation under discussion.

Let us return back to the quantum oscillator. The abstract vector space defined above has a concrete representation as follows:

$$\langle x|n \rangle = \psi_n(x) \quad \text{and} \quad \langle n|x \rangle = \psi_n^*(x).$$

Another representation is given by $\langle k|n \rangle = \psi_n(k)$, which is the Fourier transform of $\psi_n(x)$, already discussed above.

Using these new notations in terms of the abstract Hilbert vector space we can rewrite the formulas that we have encountered above. Namely:

$$\hat{a}\,|n\rangle = \sqrt{n}\,|n-1\rangle \quad \text{and} \quad \hat{a}^+\,|n\rangle = \sqrt{n+1}\,|n+1\rangle,$$

where

$$|n\rangle = N_n \left(a^+\right)^n |0\rangle, \quad \text{and} \quad \hat{a}\,|0\rangle = 0.$$

Then, e.g. the normalization condition can be expressed as follows:

$$1 = N_n^2\, \langle 0|\hat{a}^n\,(\hat{a}^+)^n|0\rangle = \langle n|n\rangle = \int\limits_{-\infty}^{+\infty} dx\, \langle n|x\rangle\langle x|n\rangle = \int\limits_{-\infty}^{+\infty} dx\, |\psi_n(x)|^2.$$

Furthermore, the orthogonality of the $|n\rangle$ states can be shown as follows:

$$\langle n|m\rangle = N_n N_m\, \langle 0|\,\hat{a}^n\,\left(\hat{a}^+\right)^m|0\rangle = 0, \quad \text{for} \quad n \neq m.$$

In fact, if $n = 1$ and $m = 2$ or $n = 2$ and $m = 1$ this can be shown explicitly by exchanging the positions of $\hat{a}$ and $\hat{a}^+$ in this expression using the commutation relations of the Heisenberg algebra. Then for the general case when $n \neq m$ this fact can be shown by induction.

Finally, the completeness condition (4.13) can also be expressed as the consequence of the following resolution of the unit operator:

$$\sum_{n=0}^{+\infty} |n\rangle \langle n| = \hat{\mathbf{1}}.$$

In fact, multiplying it's both sides by $\langle x|$ and $|y\rangle$ from the left and right, correspondingly, one obtains the relation (4.13), if uses that $\langle x|\hat{\mathbf{1}}|y\rangle = \langle x|y\rangle = \delta(x - y)$ and $\psi_n(x) \equiv \langle x|n\rangle$.

Chapter 5
Bessel Functions

Abstract This section is recorded by MIPT student Anokhin Andrei. It contains the derivation of various properties of solutions of the Bessel equation and their application to the representation theory and to the fundamental theoretical physics. At the end of this section we describe how one can represent various types of Green functions of the Klein–Gordon and Helmholtz equations in different dimensions in terms of solutions of the Bessel equation.

Consider the so called three-dimensional d'Alembert equation:

$$\left(\partial_t^2 - \Delta_2\right) \overline{f} = 0.$$

Here $\partial_t \equiv \partial/\partial t$ and $\Delta_2 \equiv \partial_x^2 + \partial_y^2$, where $\partial_x \equiv \partial/\partial x$ and $\partial_y \equiv \partial/\partial y$. We will frequently use similar notations below.

Let us look for a solution of this equation in the following form $\overline{f}(t, x, y) = e^{ikt} f_k(x, y)$, then $f_k(x, y)$ obeys the so called Helmholtz equation:

$$\left(\Delta_2 + k^2\right) f_k = 0.$$

In polar coordinates $(x, y) = (r \cos\varphi, r \sin\varphi)$ it acquires the following form:

$$\left[\frac{1}{r} \partial_r r \partial_r + \frac{1}{r^2} \partial_\varphi^2 + k^2\right] f_k = 0.$$

We define $f_k(r, \varphi) = u_n(kr) e^{in\varphi}$ and assume that $f_k(r, \varphi)$ is periodic in φ, i.e. $n \in \mathbb{Z}$ and, hence, the function f_k is well behaving in polar coordinates. Then defining $kr = z$ one can find that the function $u_n(z)$ obeys:

$$\left[\partial_z^2 + \frac{1}{z}\partial_z + \left(1 - \frac{n^2}{z^2}\right)\right] u_n = 0. \tag{5.1}$$

This is the so called Bessel equation in its basic form. This equation has two obvious peculiar points—at $z = 0$ and $z = \infty$. Let us find the asymptotic behavior of $u_n(z)$ in the vicinity of these points.

© The Author(s), under exclusive license to Springer Nature Switzerland AG 2019   41
V. Akhmedova and E. T. Akhmedov, *Selected Special Functions*
for Fundamental Physics, SpringerBriefs in Physics,
https://doi.org/10.1007/978-3-030-35089-5_5

As $z \to 0$, the Bessel equation simplifies to:

$$u_n'' + \frac{1}{z} u_n' - \frac{n^2}{z^2} u_n \approx 0,$$

which is the homogeneous in z equation. Hence, it's solution has to have the form $u_n(z) \propto z^\alpha$, for some α. In fact, substituting this ansatz into the last equation, we obtain the algebraic relation for α:

$$\alpha(\alpha - 1) + \alpha - n^2 = 0.$$

It has two solutions—$\alpha = \pm n$. And the second order differential equation under consideration has two dimensional space of solutions, which behave as:

$$u_n \approx A_{\pm n}\, z^{\pm n}, \quad \text{when} \quad z \to 0.$$

Here $A_{\pm n}$ are some constants.

At the same time when $z \to \infty$ the Bessel equation reduces to

$$\left(\partial_z^2 + 1\right) u_n(z) \approx 0.$$

Hence it's solutions behave as:

$$u_n \approx B_{\pm n}\, e^{\pm i z + \text{corrections}}, \quad \text{when} \quad z \to \infty.$$

Here $B_{\pm n}$ are some constants.

The corrections designated in the last expression will be found below. Now we will just show the corrected version:

$$u_n \approx \frac{B_{\pm n}}{\sqrt{z}}\, e^{\pm i z}, \quad \text{as} \quad z \to \infty.$$

The solution that behaves as $J_n(z) \approx \frac{1}{n!} \left(\frac{z}{2}\right)^n$, when $z \to 0$, is referred to as the Bessel function. At the same time the solution that behaves as $H_n(z) \approx \frac{e^{iz}}{\sqrt{z}}$, when $z \to \infty$, is referred to as the Hankel function.

5.1 Generating Function

As in the case of Hermite polynomials there is a convenient in applications generating function of the Bessel functions. Namely, consider the following function:

$$e^{\frac{z}{2}\left(t - \frac{1}{t}\right)} = \sum_{n=-\infty}^{+\infty} u_n(z) t^n. \tag{5.2}$$

By the direct substitution it is not hard to see that it obeys the following relation:

$$\left[\left(\partial_z^2 + \frac{1}{z}\partial_z + 1\right) - \frac{1}{z^2}\left(t^2\partial_t^2 + t\partial_t\right)\right] e^{\frac{z}{2}\left(t-\frac{1}{t}\right)} = 0.$$

Then, substituting into this equation the series expansion (5.2) and using the obvious relation:

$$(t^2\partial_t^2 + t\partial_t)\, t^n = n^2\, t^n,$$

we find that u_n from (5.2) obeys the Bessel equation (5.1). Checking the behavior of u_n as $z \to 0$ from (5.2), we find that $u_n(z) \equiv J_n(z)$, where J_n was defined at the end of the previous subsection. (This point will become clear from the discussion in the next subsection.) Thus, we have that

$$\boxed{e^{\frac{z}{2}\left(t-\frac{1}{t}\right)} = \sum_{n=-\infty}^{+\infty} J_n(z)\, t^n}$$

is the generating function of the Bessel functions.

5.2 Series Expansion

Using the generating function, we find:

$$\sum_{m=-\infty}^{+\infty} J_m(z)t^m = e^{\frac{z}{2}t}e^{-\frac{z}{2t}} = \sum_{n=0}^{\infty} \frac{\left(\frac{z}{2}\right)^n t^n}{n!} \cdot \sum_{k=0}^{\infty} \frac{\left(-\frac{z}{2}\right)^k t^{-k}}{k!}.$$

For $m > 0$ we equate the coefficients of t^m for each m separately on the both sides of this relation:

$$J_m(z) = \sum_{n-k=m} \frac{\left(\frac{z}{2}\right)^n \cdot \left(-\frac{z}{2}\right)^k}{n!\, k!} = \sum_{k=0}^{\infty} \frac{\left(\frac{z}{2}\right)^{m+k} \cdot \left(-\frac{z}{2}\right)^k}{(m+k)!\, k!} = \sum_{k=0}^{\infty} \frac{i^{2k}\, z^{2k+m}}{2^{2k+m}\, k!\,(m+k)!}.$$

Thus,

$$\boxed{J_m(z) = \sum_{k=0}^{\infty} \frac{(-1)^k}{k!\,(m+k)!} \left(\frac{z}{2}\right)^{2k+m}}, \tag{5.3}$$

for $m > 0$. Obviously this equation provides the expansion of $J_m(z)$ around $z = 0$.

5.3 Bessel Function $J_\nu(z)$ with Complex Index $\nu \in \mathbb{C}$

Consider the generalization of (5.3) which has the following form:

$$J_\nu(z) = \sum_{k=0}^{\infty} \frac{(-1)^k}{\Gamma(k+1)\,\Gamma(k+\nu+1)} \left(\frac{z}{2}\right)^{2k+\nu}, \qquad (5.4)$$

where now $\nu \in \mathbb{C}$. For $\nu = m$ it reduces to $J_m(z)$ from (5.3). At the same time, for $\nu = -m$ the first m terms of the last series vanish because of the poles of the Γ-functions in the denominator. Then, the resulting expression reduces to:

$$J_{-m}(z) = \sum_{k=m}^{\infty} \frac{(-1)^k \cdot \left(\frac{z}{2}\right)^{-m+2k}}{k!\,(k-m)!} = \sum_{l=0}^{\infty} \frac{(-1)^{l+m}\left(\frac{z}{2}\right)^{m+2l}}{(l+m)!\,l!} = (-1)^m J_m(z), \quad m = 1,\,2,\,\dots\,,$$

where on the second step we have changed the enumeration index in an obvious way. Thus, we obtain the following relation

$$J_{-m}(z) = (-1)^m J_m(z).$$

It is straightforward to show that the function $J_\nu(z)$ obeys the Bessel equation of the form:

$$u_\nu'' + \frac{1}{z} u_\nu' + \left(1 - \frac{\nu^2}{z^2}\right) u_\nu = 0. \qquad (5.5)$$

In fact, if one substitutes into this equation the series (5.4), he obtains the relation:

$$\sum_{k=0}^{\infty} z^{\nu+2k} \left[4 \cdot \alpha_{k+1}(k+1)(\nu+k+1) + \alpha_k\right] = 0,$$

that follows as the corollary of the equation $\Gamma(z+1) = z\,\Gamma(z)$ if

$$\alpha_k \equiv \frac{(-1)^k}{2^{\nu+2k}\,\Gamma(k+1)\,\Gamma(k+\nu+1)},$$

which, in their own right, coincide with the coefficients in the series (5.4).

5.4 Recurrence Relations for $J_\nu(z)$

As in the case of Hermite polynomials there are also recurrence relations for the Bessel functions. In fact, consider

$$\frac{d}{dz} z^\nu J_\nu(z) = \sum_{k=0}^{\infty} \frac{(-1)^k (2\nu + 2k)}{2^{\nu+2k} \, \Gamma(k+1) \, \Gamma(k+\nu+1)} \, z^{2\nu+2k-1}$$

$$= z^\nu \sum_{k=0}^{\infty} \frac{(-1)^k}{\Gamma(k+1)\,\Gamma(k+\nu)} \left(\frac{z}{2}\right)^{\nu-1+2k} = z^\nu J_{\nu-1}(z).$$

Thus, we have that:

$$\frac{d}{dz} z^\nu J_\nu(z) = z^\nu \, J_{\nu-1}(z).$$

Similarly, one can prove that:

$$\frac{d}{dz} z^{-\nu} J_\nu(z) = -z^{-\nu} \, J_{\nu+1}(z).$$

Performing differentiation in both recursion relations, we obtain that:

$$\boxed{J'_\nu(z) + \frac{\nu}{z} \, J_\nu(z) = J_{\nu-1}(z)} \tag{5.6}$$

and

$$\boxed{J'_\nu(z) - \frac{\nu}{z} \, J_\nu(z) = -J_{\nu+1}(z).} \tag{5.7}$$

From here we can find that:

$$\frac{2\nu}{z} J_\nu(z) = J_{\nu-1}(z) + J_{\nu+1}(z)$$

and

$$2 \, J'_\nu(z) = J_{\nu-1}(z) - J_{\nu+1}(z).$$

Exercise: Is it possible to find the same relations for $J_m(z), m \in \mathbb{Z}$ from the generating function?

5.5 Bessel Function of the Second Kind

The Bessel equation (5.5) has two solutions, which behave as z^ν and $z^{-\nu}$, when $z \to 0$. Because (5.5) is the second order differential equation one can choose the two-dimensional basis of its solutions as:

$$J_\nu(z) = \frac{\left(\frac{z}{2}\right)^\nu}{\Gamma(1+\nu)} + \dots \quad \text{and} \quad J_{-\nu}(z) = \frac{\left(\frac{z}{2}\right)^{-\nu}}{\Gamma(1-\nu)} + \dots, \quad \text{when } z \to 0.$$

Hence, a generic solution of (5.5) has the form:

$$u_\nu(x) = C_1 J_\nu(x) + C_2 J_{-\nu}(x),$$

for $\nu \notin \mathbb{Z}$. Here C_1 and C_2 are some constants.

If, however, $\nu = m \in \mathbb{Z}$, then, $J_{-m}(z) = (-1)^m J_m(z)$, as was shown above. I.e. these two solutions are not linearly independent. Hence, the above $u_\nu(z)$ is not the most general solution if $\nu \in \mathbb{Z}$. To obtain the solution, which is applicable even for $\nu \in \mathbb{Z}$ let us introduce the following function:

$$Y_\nu(z) = \frac{J_\nu(z) \cos(\pi \nu) - J_{-\nu}(z)}{\sin(\pi \nu)}.$$

Then one can define:

$$Y_m(z) = \lim_{\nu \to m} Y_\nu(z).$$

According to the l'Hopital's rule:

$$Y_m(z) = \frac{1}{\pi} \left(\left. \frac{\partial J_\nu}{\partial \nu} \right|_{\nu=m} - (-1)^m \left. \frac{\partial J_{-\nu}}{\partial \nu} \right|_{\nu=m} \right).$$

Thus, the general solution of the Bessel equation (5.5) for all values of ν is as follows:

$$u_\nu(x) = C_1 J_\nu(x) + C_2 Y_\nu(x),$$

where C_1 and C_2 are some complex constants.

The function $Y_\nu(z)$ obeys the same recursion relations as $J_\nu(z)$. This fact straightforwardly follows from its definition provided above. Also, it is not hard to see that $Y_{-m}(z) = (-1)^m Y_m(z)$ as follows from its definition.

5.6 Series Expansion for $Y_m(z)$

Let us find the series expansion for $Y_m(z)$, which is similar to (5.3). First:

$$\left[\frac{\partial J_\nu}{\partial \nu} \right]\Big|_{\nu=m} = \sum_{k=0}^{\infty} \frac{(-1)^k \cdot \left(\frac{z}{2}\right)^{m+2k}}{k! \, (k+m)!} \cdot \left\{ \log \frac{z}{2} - \psi(k+m+1) \right\}, \quad \text{where } \psi(z) \equiv \frac{\Gamma'(z)}{\Gamma(z)}.$$

Second:

$$\left[\frac{\partial J_{-\nu}}{\partial \nu} \right]\Big|_{\nu=m} = \sum_{k=0}^{\infty} \frac{(-1)^k \cdot \left(\frac{z}{2}\right)^{2k-\nu}}{k! \, \Gamma(k-\nu+1)} \cdot \left\{ -\log \frac{z}{2} + \psi(k-\nu+1) \right\}.$$

For $k = 0, 1, \ldots, (m-1)$ we have that $\Gamma(k - \nu + 1) \to \infty$ and $\psi(k - \nu + 1) \to \infty$, as $\nu \to m$. Hence, first m terms in the last series are not defined. However, one can deduce that:

$$\psi(1 - z) - \psi(z) = \pi \cot(\pi z),$$

which follows from the logarithmic differential of the relation (2.2).

Also using that $\Gamma(n + 1) = n!$ we can proceed further. In fact, for $k = 0, 1, \ldots, (m-1)$ we can find that:

$$\lim_{\nu \to m} \frac{\psi(k - \nu + 1)}{\Gamma(k - \nu + 1)} = \lim_{\nu \to m} \left\{ \Gamma(\nu - k) \sin[\pi(\nu - k)] \frac{\psi(\nu - k) + \pi \cot[\pi(\nu - k)]}{\pi} \right\} =$$

$$= (-1)^{m-k}(m - k - 1)!.$$

As the result, one obtains that:

$$\left[\frac{\partial J_{-\nu}}{\partial \nu} \right]\Big|_{\nu=m} = (-1)^m \sum_{k=0}^{m-1} \frac{(m - k - 1)!}{k!} \left(\frac{z}{2} \right)^{2k-m} +$$

$$+ (-1)^m \sum_{p=0}^{\infty} \frac{(-1)^p}{(m + p)! \, p!} \left\{ -\log \frac{z}{2} + \psi(p + 1) \right\} \left(\frac{z}{2} \right)^{2p+m}.$$

Hence:

$$\boxed{\begin{aligned} Y_m(z) = &-\frac{1}{\pi} \sum_{k=0}^{m-1} \frac{(m - k - 1)!}{k!} \left(\frac{z}{2} \right)^{2k-m} + \\ &+ \frac{1}{\pi} \sum_{k=0}^{\infty} \frac{(-1)^k \cdot \left(\frac{z}{2} \right)^{2k+m}}{k! \, (m + k)!} \left\{ 2 \log \frac{z}{2} - \psi(k + 1) - \psi(k + m + 1) \right\}. \end{aligned}}$$

When $m = 0$ the first sum is just equal to zero.

5.7 Hankel and MacDonald Functions

Another basis of solutions of the Bessel equation (5.5) is given by the so called Hankel functions of the first kind:

$$\boxed{H_\nu^{(1)}(z) \equiv J_\nu(z) + i Y_\nu(z) = \frac{J_{-\nu}(z) - e^{-i\pi\nu} J_\nu(z)}{i \sin(\pi \nu)},}$$

and of the second kind:

$$H_\nu^{(2)}(z) \equiv J_\nu(z) - iY_\nu(z) = \frac{e^{i\pi\nu} J_\nu(z) - J_{-\nu}(z)}{i \sin(\pi\nu)}.$$

Furthermore, consider the following functions:

$$I_\nu(z) = \sum_{k=0}^{\infty} \frac{\left(\frac{z}{2}\right)^{\nu+2k}}{\Gamma(k+1)\,\Gamma(k+\nu+1)},$$

and

$$K_\nu(z) = \frac{\pi}{2} \cdot \frac{I_{-\nu}(z) - I_\nu(z)}{\sin(\pi\nu)}.$$

Also one can define $K_m(z) = \lim_{\nu\to m} K_\nu(z)$.

It is not hard to see that:

$$I_\nu(z) = J_\nu(iz)e^{-\frac{i\pi\nu}{2}},$$

and

$$K_\nu(z) = \frac{i\pi}{2} e^{\frac{i\pi\nu}{2}} H_\nu^{(1)}(iz) = -\frac{i\pi}{2} e^{-\frac{i\pi\nu}{2}} H_\nu^{(2)}(-iz).$$

These are the so called MacDonald or modified Bessel functions. From the last relations one can deduce that they obey the following equation:

$$u_\nu'' + \frac{1}{z} u_\nu' - \left(1 + \frac{\nu^2}{z^2}\right) u_\nu = 0,$$

which can be obtained from the Bessel equation by the change $z \to i\,z$.

It is straightforward to see that for the Hankel and MacDonald functions we have the same recursion relations as for $J_\nu(z)$.

5.8   Bessel, Hankel and MacDonald Functions of Half-Integer Indexes

Consider the J_ν Bessel function with the index $\nu = 1/2$:

$$J_{\frac{1}{2}}(z) = \sum_{k=0}^{\infty} \frac{(-1)^k \left(\frac{z}{2}\right)^{\frac{1}{2}+2k}}{\Gamma(k+1)\,\Gamma(k+\frac{3}{2})} = \left(\frac{2z}{\pi}\right)^{1/2} \sum_{k=0}^{\infty} \frac{(-1)^k z^{2k}}{\Gamma(2k+2)} = \left(\frac{2}{\pi z}\right)^{1/2} \sin(z).$$

where we have used the relation (2.3). Similarly one can show that:

$$J_{-\frac{1}{2}}(z) = \left(\frac{2}{\pi z}\right)^{1/2} \cos(z).$$

Then, from the recursion relations it follows that:

$$J_{\frac{3}{2}}(z) = \frac{1}{z} J_{\frac{1}{2}}(z) - J_{-\frac{1}{2}}(z) = \left(\frac{2}{\pi z}\right)^{1/2} \left[\frac{\sin z}{z} - \cos z\right],$$

and

$$J_{-\frac{3}{2}}(z) = -\left(\frac{2}{\pi z}\right)^{1/2} \left[\sin z + \frac{\cos z}{z}\right].$$

Furthermore, it is not hard to deduce the following relations:

$$Y_{\frac{1}{2}}(z) = -J_{-\frac{1}{2}}(z) = -\left(\frac{2}{\pi z}\right)^{1/2} \cos z$$

$$\text{and} \quad H_{\frac{1}{2}}^{(1)} = -i \left(\frac{2}{\pi z}\right)^{1/2} e^{iz} = H_{\frac{1}{2}}^{(2)*}(z),$$

$$I_{\frac{1}{2}}(z) = \left(\frac{2}{\pi z}\right)^{1/2} \sinh z, \quad I_{-\frac{1}{2}}(z) = \left(\frac{2}{\pi z}\right)^{1/2} \cosh z$$

$$\text{and} \quad K_{\frac{1}{2}}(z) = \left(\frac{2}{\pi z}\right)^{1/2} e^{-z} \frac{\pi}{2} = \left(\frac{\pi}{2z}\right)^{1/2} e^{-z}.$$

Thus, the solutions of the Bessel equation for half integer indexes can be expressed via the standard trigonometric and exponential functions.

5.9 Integral Representation

Let us derive the integral representation of the Bessel functions. For that we will use the integral representation of the Γ-function (2.7) for the case when $z = k + \nu + 1$. Then, we can represent the Bessel function as:

$$J_\nu(z) = \sum_{k=0}^{\infty} \frac{(-1)^k \cdot \left(\frac{z}{2}\right)^{\nu+2k}}{\Gamma(k+1)} \cdot \frac{1}{2\pi i} \int_{C^*} e^s s^{-(k+\nu+1)} ds =$$

$$= \left(\frac{z}{2}\right)^\nu \frac{1}{2\pi i} \int_{C^*} e^s s^{-\nu-1} ds \sum_{k=0}^{\infty} \frac{(-1)^k \left(\frac{z^2}{4s}\right)^k}{\Gamma(k+1)} = \left(\frac{z}{2}\right)^\nu \frac{1}{2\pi i} \int_{C^*} e^{s-\frac{z^2}{4s}} s^{-\nu-1} ds.$$

Hence, we obtain that:

$$J_\nu(z) = \frac{1}{2\pi i} \int\limits_{C^*} e^{\frac{z}{2}\left(s - \frac{1}{s}\right)} s^{-\nu-1} ds, \text{ where } |\arg z| < \frac{\pi}{2}.$$

Let Re $z > 0$, which is the same as the condition $|\arg z| < \frac{\pi}{2}$, and make the change of variables as $s = e^{i\xi}$ in the last expression, then the contour C^* transforms into C, which is illustrated on the figure:

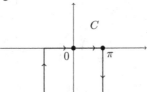

As the result we obtain another form of the integral representation:

$$J_\nu(z) = \frac{1}{2\pi} \int\limits_{C} e^{iz\sin\xi - i\nu\xi} \, d\xi.$$

Exercise: _Changing $t = e^{i\theta}$ in the generating function for $J_m(z)$, $m \in \mathbb{Z}$, find from it an integral representation for $J_m(z)$._

To obtain the integral representation for $H_\nu^{(1,\,2)}(z)$, let us use the following trick. Rewrite the Bessel equation in the following form:

$$z^2 u_\nu'' + z u_\nu' + (z^2 - \nu^2) u_\nu = 0.$$

Then represent it's solution as:

$$u_\nu(z) = \int\limits_{C} \mathcal{K}(z, \xi) \mathcal{W}(\xi) d\xi,$$

where $\mathcal{W}(\xi)$, C and $\mathcal{K}(z, \xi)$ we will specify in a moment. Plaguing the last expression into the Bessel equation, we obtain that:

$$0 = \int\limits_{C} \left[z^2 \partial_z^2 \mathcal{K} + z \partial_z \mathcal{K} + (z^2 - \nu^2)\mathcal{K}\right] \mathcal{W}(\xi) d\xi.$$

Let us define $\mathcal{K}$ in such a way that it obeys the equation:

$$\left[z^2 \partial_z^2 + z \partial_z + z^2 + \partial_\xi^2\right] \mathcal{K} = 0. \tag{5.8}$$

Then, we have that:

$$\int_C \left[\partial_\xi^2 \mathcal{K} + v^2 \mathcal{K} \right] \mathcal{W}(\xi) d\xi = 0.$$

Integrating here the first term by parts two times, we find that:

$$\int_C \left[\mathcal{W}'' + v^2 \mathcal{W} \right] \mathcal{K} d\xi + \left[\mathcal{W} \partial_\xi \mathcal{K} - \mathcal{K} \mathcal{W}' \right] \Big|_a^b = 0,$$

where a and b are the ends of the contour C. Thus, if $\mathcal{W} = e^{\pm iv\xi}$ and if we choose the integration contour C in such a way that the expression $\mathcal{W} \partial_\xi \mathcal{K} - \mathcal{K} \mathcal{W}'$ is vanishing at its ends a and b, then the above defined function $u_v(z)$ does solve the Bessel equation. At the same time it is not hard to find that $\mathcal{K}(\xi, z) = e^{iz \sin \xi}$ solves Eq. (5.8).

Furthermore, as the contour C we can choose either C_1 or C_2, which are illustrated on the figure:

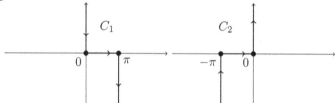

On the imaginary axis in the complex ξ-plane we have that $\sin \xi = \sin i\eta = i \sinh \eta$, $\eta \in \mathbb{R}$. At the same time on the axis where $\xi = \pm \pi + i\eta$, $\eta \in \mathbb{R}$, we have that $\sin \xi = -\sin i\eta = -i \sinh \eta$. Hence,

$$|\mathcal{K}(z, \xi)| = \begin{cases} e^{-x \sinh \eta} & , \ \eta > 0 \\ e^{x \sinh \eta} & , \ \eta < 0, \end{cases}$$

where $x = \operatorname{Re} z$. Thus, if $\operatorname{Re} z > 0$ then as $\eta \to \pm\infty$ it follows that $|\mathcal{K}| \to 0$. At the same time:

$$\mathcal{W} \partial_\xi \mathcal{K} = e^{\pm iv\xi} iz \cos \xi \mathcal{K} \quad \text{and} \quad \mathcal{K} \mathcal{W}' = \pm iv e^{\pm iv\xi} \mathcal{K}$$

both also tend to zero at the ends of the both contours C_1 and C_2. Thus, we have two independent solutions of the Bessel equation:

$$H_v^{(1)}(z) = \frac{1}{\pi} \int_{C_1} e^{iz \sin \xi - iv\xi} d\xi \quad \text{and} \quad H_v^{(2)}(z) = \frac{1}{\pi} \int_{C_2} e^{iz \sin \xi - iv\xi} d\xi.$$

The fact that the contour C_1 corresponds to the Hankel function of the first kind $H^{(1)}$, while the contour C_2—to the Hankel function of the second kind, $H^{(1)}$, can be established e.g. from the series expansion of the integrals in powers of z.

Then, in fact, we also can find that:

$$H_\nu^{(1)}(z) + H_\nu^{(2)}(z) = \frac{1}{\pi} \int_C e^{iz\sin\xi - i\nu\xi}\, d\xi = 2J_\nu(z),$$

which coincides with the above defined integral representation for $J_\nu(z)$.

Yet another integral representation for $J_\nu(z)$ can be obtained as follows. Using the definition of the Euler's B-function, one an write that:

$$\frac{1}{\Gamma(k+\nu+1)} = \frac{1}{\Gamma\left(k+\frac{1}{2}\right)\Gamma\left(\nu+\frac{1}{2}\right)} \int_{-1}^{1} t^{2k}\left(1-t^2\right)^{\nu-\frac{1}{2}}\, dt, \quad \text{Re}\,\nu > -\frac{1}{2}.$$

Substituting this into the series expansion of J_ν we obtain that:

$$J_\nu(z) = \sum_{k=0}^{\infty} \frac{(-1)^k\,(z/2)^{\nu+2k}}{\Gamma(k+1)} \frac{1}{\Gamma\left(k+\frac{1}{2}\right)\Gamma\left(\nu+\frac{1}{2}\right)} \int_{-1}^{1} t^{2k}\left(1-t^2\right)^{\nu-\frac{1}{2}}\, dt =$$

$$= \frac{(z/2)^\nu}{\Gamma\left(\nu+\frac{1}{2}\right)} \int_{-1}^{1} dt\,\left(1-t^2\right)^{\nu-\frac{1}{2}} \sum_{k=0}^{\infty} \frac{(-1)^k\,(z\,t)^{2k}}{2^{2k}\,\Gamma(k+1)\,\Gamma\left(k+\frac{1}{2}\right)} =$$

$$= \frac{(z/2)^\nu}{\sqrt{\pi}\,\Gamma\left(\nu+\frac{1}{2}\right)} \int_{-1}^{1} dt\,\left(1-t^2\right)^{\nu-\frac{1}{2}} \cos(z\,t),$$

where we have used the properties of the Γ-function. Thus,

$$J_\nu(z) = \frac{(z/2)^\nu}{\sqrt{\pi}\,\Gamma\left(\nu+\frac{1}{2}\right)} \int_{-1}^{1} dt\,\left(1-t^2\right)^{\nu-\frac{1}{2}} \cos(z\,t), \quad \text{Re}\,\nu > -\frac{1}{2}, \quad |\arg z| < \pi. \qquad (5.9)$$

The restriction $|\arg z| < \pi$ here means that there is the cut in the complex plane going along the negative real axis.

5.10 Asymptotic Form for the Large Argument

Using the steepest decent or the stationary phase approximation of the above integrals, we can find that as $|z| \gg \nu$:

$$H_\nu^{(1,\,2)}(z) \approx \sqrt{\frac{2}{\pi z}}\, e^{\pm i\left(z - \frac{\pi}{2}\nu - \frac{\pi}{4}\right)},$$

$$J_\nu(z) \approx \sqrt{\frac{2}{\pi z}}\, \cos\left(z - \frac{\pi}{2}\nu - \frac{\pi}{4}\right) \quad \text{and} \quad Y_\nu(z) \approx \sqrt{\frac{2}{\pi z}}\, \sin\left(z - \frac{\pi}{2}\nu - \frac{\pi}{4}\right).$$

Note for future reference that from the last expression one can see that each Bessel function has infinite number of zeros.

Furthermore,

$$I_\nu(z) \equiv e^{-\frac{i\pi\nu}{2}} J_\nu(i\,z) \approx \frac{1}{\sqrt{2\pi z}} e^z,$$

and

$$K_\nu(z) \equiv \frac{\pi i}{2} e^{\frac{i\pi\nu}{2}} H_\nu^{(1)}(i\,z) \approx \sqrt{\frac{\pi}{2z}} e^{-z},$$

in the same limit as above.

The simplest way to find all these expressions is to calculate first the expressions for $H^{(1)}$ and $H^{(2)}$ using the stationary phase or the steepest descent method, and then find all the other functions via the relations that have been established in the Sect. 5.7.

5.11 Orthogonality

Consider the following function $f(\lambda, z) = \sqrt{z}\, J_n(\lambda z)$. It is straightforward to see that is solves the following equation:

$$\left[\frac{d^2}{dz^2} - \frac{n^2 - \frac{1}{4}}{z^2} \right] f(\lambda, z) = -\lambda^2 f(\lambda, z).$$

As a consequence of this equation it is possible to deduce the relation as follows:

$$f(\lambda, z)\, f''(\mu, z) - f''(\lambda, z)\, f(\mu, z) = (\lambda^2 - \mu^2) f(\lambda, z) f(\mu, z).$$

To find it one just has to multiply the above equation by $f(\mu, z)$ and subtract from it the equation for $f(\mu, z)$ multiplied by $f(\lambda, z)$. Here the prime means the differentiation with respect to z.

Integrating the relation under consideration over $z \in [0, 1]$, we obtain:

$$\int_0^1 f(\lambda, z) f(\nu, z)\, dz = \frac{\left[f(\lambda, 1) f'(\mu, 1) - f'(\lambda, 1) f(\mu, 1) \right]}{\lambda^2 - \mu^2}.$$

Thus,

$$\int_0^1 J_n(\lambda z) J_n(\mu z)\, z\, dz = \frac{J_n(\lambda)\, \mu\, J_n'(\mu) - J_n(\mu)\, \lambda\, J_n'(\lambda)}{\lambda^2 - \mu^2}.$$

Using the recursion relations for the Bessel functions, that we have derived above, we can rewrite this equation as:

$$\int_0^1 J_n(\lambda z) J_n(\mu z) z\, dz = \frac{-J_n(\lambda)\, \mu\, J_{n+1}(\mu) + J_n(\mu)\, \lambda\, J_{n+1}(\lambda)}{\lambda^2 - \mu^2}. \tag{5.10}$$

Then taking the limit $\mu \to \lambda$ and using the l'Hopital's rule, one obtains:

$$\int_0^1 J_n^2(\lambda z) z\, dz = \frac{1}{2}\left[J_{n+1}^2(\lambda) + J_n^2(\lambda) - \frac{2n}{\lambda} J_n(\lambda) J_{n+1}(\lambda) \right], \tag{5.11}$$

where we have used the recursion relations again.

Let us consider such functions $f(\lambda, z)$ that $f(\lambda, z = 1) = 0$ and that $f(\lambda, z = 0)$ is regular. Then, from the collection $f(\lambda, z)$ for all λ it is appropriate to choose function with $\lambda = \gamma_k$, where $J_n(\gamma_k) = 0$, i.e. γ_k are zeros of $J_n(x)$. Thus, we define the following infinite chain of functions enumerated by

$$\psi_k(z) \equiv \sqrt{z}\, J_n(\gamma_k z).$$

The chain is infinite because there is infinite number of zeros of J_n as we have explained above.

Then, as follows from (5.10), we obtain the orthogonality condition:

$$\int_0^1 \psi_k(z)\, \psi_m(z)\, dz = 0,$$

if $k \neq m$. The normalization of $\psi_k(z)$ we can find from Eq. (5.11):

$$\int_0^1 J_n^2(\gamma_k z)\, z\, dz = \frac{1}{2} J_{n+1}^2(\gamma_k).$$

Thus, $\psi_k(z)$, $k \in \mathbb{N}$ can be used as a basis of such functions on the interval $z \in [0, 1]$ that vanish at $z = 1$.

5.12 Addition or Summation Theorems for $J_m(z)$

Bessel functions $J_m(z)$ obey an important relation, which we will derive in this subsection. Consider the following sequence of equalities:

$$\sum_{n=-\infty}^{+\infty} J_n(z_1 + z_2)w^m = e^{\frac{z_1+z_2}{2}(w-\frac{1}{w})} = e^{\frac{z_1}{2}(w-\frac{1}{w})}e^{\frac{z_2}{2}(w-\frac{1}{w})}$$

$$= \sum_{m=-\infty}^{+\infty} J_m(z_1)w^m \sum_{k=-\infty}^{+\infty} J_k(z_2)w^k,$$

where we have used the expression for the generating function.

Thus,

$$\sum_{n=-\infty}^{+\infty} J_n(z_1 + z_2)w^n = \sum_{n=-\infty}^{+\infty}\left[\sum_{k=-\infty}^{+\infty} J_k(z_1)J_{n-k}(z_2)\right]w^n.$$

Equating the expressions which multiply equal powers of w^n on both sides of this relation for each n separately, we obtain:

$$\boxed{J_n(z_1 + z_2) = \sum_{k=-\infty}^{+\infty} J_k(z_1)J_{n-k}(z_2),}$$

which is the so called summation relation for the Bessel functions. It is similar in spirit to the relation between the trigonometric functions, which follow from the simple relation $e^{z_1+z_2} = e^{z_1} e^{z_2}$.

5.13 Relation to the Group Representation Theory

As in the case of the Hermite polynomials, for the Bessel functions we also encounter a relation to a symmetry algebra. In fact, consider the so called $ISO(2)$ algebra, which is the Poincaré symmetry of the Euclidian two-dimensional plane. It is generated by the translations and rotations of the plane, as we will explain in a moment.

Namely, the generators of this group are as follows: the operator

$$\hat{a}_1 = \partial_x$$

generates translations along the x direction. In fact, for a very small (infinitesimal) α we have that

$$f(x + \alpha, y) \approx f(x, y) + \alpha\, \partial_x f(x, y).$$

Similarly the operator

$$\hat{a}_2 = \partial_y$$

generates translations along the y direction. At the same time, the operator

$$\hat{a}_3 = y\partial_x - x\partial_y$$

generates rotations. In fact, for a rotation by an infinitesimal angle ϕ the functions $f(x, y)$ transforms as $f(x, y) \rightarrow f(x + y\phi, y - x\phi)$. Then

$$f(x + y\phi, y - x\phi) \approx f(x, y) + \phi \left(y\partial_x - x\partial_y \right) f(x, y).$$

In polar coordinates $(x, y) = (r \cos \varphi, r \sin \varphi)$ these generators acquire the following form:

$$\hat{a}_1 = \cos \varphi \, \partial_r - \frac{\sin \varphi}{r} \partial_\varphi, \quad \hat{a}_2 = \sin \varphi \, \partial_r + \frac{\cos \varphi}{r} \partial_\varphi, \quad \text{and} \quad \hat{a}_3 = -\partial_\varphi.$$

Frequently one also uses the following operators $\hat{h}_\pm = \hat{a}_1 \pm i\hat{a}_2 = -e^{\pm i\varphi} \left(\partial_r + \frac{i}{r}\partial_\varphi \right)$ instead of $\hat{a}_1$ and $\hat{a}_2$.

From the above definition of the generators one can find that the algebra has the following commutation relations:

$$\left[\hat{h}_+, \hat{h}_- \right] = 0, \quad \left[\hat{h}_+, \hat{a}_3 \right] = i \, \hat{h}_+ \quad \text{and} \quad \left[\hat{h}_-, \hat{a}_3 \right] = -i \, \hat{h}_-,$$

or

$$\left[\hat{a}_1, \hat{a}_2 \right] = 0, \quad \left[\hat{a}_2, \hat{a}_3 \right] = \hat{a}_1 \quad \text{and} \quad \left[\hat{a}_3, \hat{a}_1 \right] = \hat{a}_2.$$

It is straightforward to calculate that the result of the action of these generators on the Bessel functions is as follows:

$$\hat{h}_+ \, e^{in\varphi} \, J_n(r) = e^{i(n+1)\varphi} \, J_{n+1}(r).$$

This equation follows from the recurrence relation (5.7) in which $\nu = n$ and $z = r$. Similarly, the equation:

$$\hat{h}_- \, e^{in\varphi} \, J_n(r) = e^{i(n-1)\varphi} \, J_{n-1}(r)$$

follows from the other recurrence relation (5.6). The action of the $\hat{a}_3$ operator on $e^{in\varphi} J_n(r)$ is very easy to find.

Furthermore, from these relations we obtain the equation:

$$\hat{h}_+ \, \hat{h}_- \, e^{in\varphi} \, J_n(r) = -e^{in\varphi} \, J_n(r)$$

Opening the brackets here we find the Bessel equation (5.1) for J_n, where $z = r$.

Thus, the infinite vector $\left(\ldots, J_{-1} e^{-i\varphi}, J_0, J_1 e^{i\varphi}, J_2 e^{i2\varphi}, \ldots \right)$ provides a representation of the $ISO(2)$ algebra. Moreover, while each of the generators $\hat{h}_\pm$ when acting on the vector in question is not diagonal:

$$
\hat{h}_+ \begin{pmatrix} \vdots \\ J_{-1}\,e^{-i\varphi} \\ J_0 \\ J_1\,e^{i\varphi} \\ J_2\,e^{i\,2\varphi} \\ \vdots \end{pmatrix} = \begin{pmatrix} \cdots & \vdots & \vdots & \vdots & \vdots & \cdots \\ \cdots & 0 & 1 & 0 & 0 & \cdots \\ \cdots & 0 & 0 & 1 & 0 & \cdots \\ \cdots & 0 & 0 & 0 & 1 & \cdots \\ \cdots & 0 & 0 & 0 & 0 & \cdots \\ \cdots & \vdots & \vdots & \vdots & \vdots & \cdots \end{pmatrix} \begin{pmatrix} \vdots \\ J_{-1}\,e^{-i\varphi} \\ J_0 \\ J_1\,e^{i\varphi} \\ J_2\,e^{i\,2\varphi} \\ \vdots \end{pmatrix}
$$

and

$$
\hat{h}_- \begin{pmatrix} \vdots \\ J_{-1}\,e^{-i\varphi} \\ J_0 \\ J_1\,e^{i\varphi} \\ J_2\,e^{i\,2\varphi} \\ \vdots \end{pmatrix} = \begin{pmatrix} \cdots & \vdots & \vdots & \vdots & \vdots & \cdots \\ \cdots & 0 & 0 & 0 & 0 & \cdots \\ \cdots & -1 & 0 & 0 & 0 & \cdots \\ \cdots & 0 & -1 & 0 & 0 & \cdots \\ \cdots & 0 & 0 & -1 & 0 & \cdots \\ \cdots & \vdots & \vdots & \vdots & \vdots & \cdots \end{pmatrix} \begin{pmatrix} \vdots \\ J_{-1}\,e^{-i\varphi} \\ J_0 \\ J_1\,e^{i\varphi} \\ J_2\,e^{i\,2\varphi} \\ \vdots \end{pmatrix}
$$

the matrix corresponding to their product, $\hat{h}_+\,\hat{h}_-$, appears to be diagonal. The situation is similar to the one we encounter in the case of the Hermite polynomials. Namely, while $\hat{a}$ and $\hat{a}^+$ are rotating the vector $(H_0, H_1, \dots)$, their product, $\hat{a}^+\,\hat{a}$, is keeping it intact. This is just a consequence of the fact that this vector provides solutions of the equation under consideration, i.e. it is the eigen-vector of the corresponding operator.

5.14 Application: General Discussion of the Green Functions

There are many physical problems where Bessel functions find their application. We will discuss here one of such situations, which is the Green functions for the so called Klein–Gordon equation. But before doing that let us consider the general idea of how to construct the Green functions on the simplest examples. Suppose one would like to solve the following equation

$$
\left(-\frac{d^2}{dx^2} + M^2 \right) \phi(x) = J(x), \tag{5.12}
$$

where $J(x)$ is an arbitrary given function and one has to find $\phi(x)$; M here is a constant which has the dimensionality, that is inverse to the length. We assume here that x is real.

It is not hard to see that if one knows the Green function $G(x)$, which by definition solves the equation

$$\left(-\frac{d^2}{dx^2} + M^2\right) G(x) = \delta(x),$$

then he can find $\phi(x)$ as follows:

$$\phi(x) = \int_{-\infty}^{+\infty} dx' \, G\left(x - x'\right) J\left(x'\right).$$

In fact, let us substitute this $\phi(x)$ into the Eq. (5.12). The result is:

$$\left(-\frac{d^2}{dx^2} + M^2\right) \int_{-\infty}^{+\infty} dx' \, G\left(x - x'\right) J\left(x'\right)$$
$$= \int_{-\infty}^{+\infty} dx' \left[\left(-\frac{d^2}{dx^2} + M^2\right) G\left(x - x'\right)\right] J\left(x'\right)$$
$$= \int_{-\infty}^{+\infty} dx' \, \delta\left(x - x'\right) J\left(x'\right) = J(x).$$

Thus, such a $\phi(x)$ indeed solves (5.12).

Let us find this Green function $G(x)$. After the Fourier transformation

$$G(x) = \int_{-\infty}^{\infty} \frac{dp}{2\pi} \widetilde{G}(p) \, e^{-i \, p x}$$

the equation for the Green function acquires the following form:

$$\left[p^2 + M^2\right] \widetilde{G}(p) = 1.$$

Hence,

$$G(x) = \int_{-\infty}^{\infty} \frac{dp}{2\pi} \frac{1}{M^2 + p^2} e^{-i \, p x}.$$

To take this integral one has to close the contour of integration in the complex p-plane and use the Jordan's lemma. When $x > 0$ one has to close the contour clockwise on the lower complex half p-plane, while when $x < 0$ it is necessary to close the contour counter-clockwise on the upper complex half p-plane. The closure of the contour does not add anything to the integral, because the integrand is zero on the corresponding semicircle.

Then for the case of $x > 0$ the result of integration with the use of the Cauchy theorem is as follows:

$$G(x) = \frac{1}{2M} e^{-M x},$$

while when $x < 0$ the result is:

$$G(x) = \frac{1}{2M} e^{M x}.$$

Hence, the complete solution for all values of x can be expressed as

$$G(x) = \frac{1}{2M} e^{-M|x|}.$$

Note that the homogeneous equation (5.12) (for the case when $J(x) = 0$) does have only exponential solutions $\phi_0(x) = e^{\pm Mx}$, which grow indefinitely as either $x \to +\infty$ or $x \to -\infty$. As the result, the inhomogeneous equation for the Green function has a unique solution which we have just found.

Consider now the Green function for a little bit different operator:

$$\left[\frac{d^2}{dx^2} + M^2\right] G(x) = \delta(x). \tag{5.13}$$

We can call this as the one-dimensional Klein–Gordon equation. The homogeneous form of it:

$$\left[\frac{d^2}{dx^2} + M^2\right] \phi_0(x) = 0,$$

has everywhere finite and normalizable solutions $\phi_0(x) = e^{\pm iMx}$. As the result, unlike the previous case, the equation for the Green function under consideration does not have a unique solution.

Let us have a closer look at this equation. Again after the Fourier transformation it is necessary to take the following integral:

$$G(x) = \int_C \frac{dp}{2\pi} \frac{1}{M^2 - p^2} e^{-ipx}, \tag{5.14}$$

where the choice of the contour C is exactly the issue here. For any choice of the contour C (with the appropriate asymptotics) in the complex p-plane this integral solves the equation under consideration. And different choices of the contour provide solutions which differ from each other by additions of solutions ϕ_0 of the homogeneous equation.

There are four principally different ways to draw the contour C in the complex p-plane, as is shown on the figure:

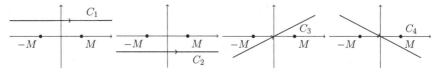

As the result we have the following four different Green functions:

1. Retarded Green function $G_R(x)$.
2. Advanced Green function $G_A(x)$.
3. Feynman time-ordered propagator $G_F(x)$.
4. Feynman anti-time-ordered propagator $G_{\overline{F}}(x)$.

We have to close the contour C in the complex p-plane according to the same rules as it was explained above in this subsection. Then the retarded Green function, corresponding to the contour C_1, is as follows

$$G_R(x) = \theta(x) \frac{1}{M} \sin(M x),$$

where $\theta(x)$ is the Heaviside's step function. This Green function is referred to as retarded because $G_R\left(x - x'\right)$ is not zero only when $x > x'$. In fact, consider x as a time variable. Then, while $\delta\left(x - x'\right)$ defines a source at a moment x' (in Eq. (5.13) $x' = 0$), the function $G_R\left(x - x'\right)$ is not zero only if the event x happens after the event x'.

The advanced Green function G_A corresponds to the contour C_2 and is equal to:

$$G_A(x) = -\theta(-x) \frac{1}{M} \sin(M x).$$

It is referred to as advanced for the same reason why G_R is referred to as retarded.

The Feynman propagator corresponds to the contour C_3 and is equal to:

$$i G_F(x) = \frac{i}{2M} e^{-i M |x|}$$

And, finally, the anti-time-ordered propagator corresponds to the contour C_4 and is equal to:

$$i G_{\overline{F}}(x) = -\frac{i}{2M} e^{i M |x|}$$

It is not hard to see that the differences between these four Green functions solve the homogeneous form of the equation under consideration.

Note that also instead of shifting the contours C_1, C_2, C_3 and C_4 from the real line one can shift the poles of the integrand in the Fourier transformation. Then the integral in (5.14) goes over the real axis, while the poles are shifted according to the following rules:

- The retarded Green function corresponds to the shift as follows: $p^2 - M^2 \to p^2 - M^2 - i\, 0\, \text{sign}(p)$;
- The advanced Green function corresponds to the shift $p^2 - M^2 \to p^2 - M^2 + i\, 0\, \text{sign}(p)$;
- The Feynman propagator corresponds to $p^2 - M^2 \to p^2 - M^2 + i\, 0$;
- The anti-time-ordered Feynman propagator corresponds to $p^2 - M^2 \to p^2 - M^2 - i\, 0$.

And, finally, it is also not hard to see that the Feynman propagator and the Green function for the Eq. (5.12) are equal to each other, if considered as complex functions in the complex plane of the distance $l = |x|$. In fact, if we consider the function

$G(l) = \frac{1}{2M} e^{-Ml}$, then for real l it coincides with the Green function of Eq. (5.12), while for pure imaginary l this function is equal to the Feynman propagator.[1]

Moreover, it is not hard to see that $G_R(l) = \theta(x) \operatorname{Im} G_F(l)$.

5.15 Application: Green Functions of the Klein–Gordon Equation

Let us apply the gained in the previous subsection knowledge to the case of two-dimensional Green functions. Consider, first, the generally covariant action for the real scalar field:

$$S = \int d^D \underline{x} \sqrt{|g|} \left[g^{\mu\nu} \partial_\mu \phi \, \partial_\mu \phi - M^2 \phi^2 \right], \quad \underline{x} = x_\mu = (x_0, \ldots, x_{D-1}).$$

Here we use the units in which the Planck constant and the speed of light are set to one: $\hbar = 1 = c$; M has the dimension of energy and is referred to as the mass of the field. In the units under consideration the dimension of energy coincides with the inverse dimension of length.

At the same time $g = \det g_{\mu\nu}$, where $g_{\mu\nu}$ is the metric tensor of the D-dimensional space or space-time: $ds^2 = g_{\mu\nu} dx^\mu dx^\nu$, $\mu, \nu = \overline{0, D-1}$; $g^{\mu\nu}$ is the inverse metric tensor $g_{\mu\nu} g^{\nu\alpha} = \delta^\alpha_\mu$. Below, in the case if we deal with the space-time, the signature of the metric is as follows: $(+, -, -, \ldots)$.

From the least action principle we obtain the following equations of motion for the scalar field:

$$\left[\Box(g) + M^2 \right] \phi = 0, \quad \text{where} \quad \Box(g) \equiv \frac{1}{\sqrt{|g|}} \partial_\mu g^{\mu\nu} \sqrt{|g|} \, \partial_\nu. \tag{5.15}$$

The sign $\Box(g)$ here designates either the Laplace or d'Alambert operator acting on the scalar fields depending on whether we deal with Euclidian space or Minkowskian space-time. We will use this definition of $\Box(g)$ below.

In this subsection we are interested in the Feynman Green function or propagator of the so called Klein–Gordon operator, which is defined as:

$$\left[\Box(g) + M^2 \right] G(\underline{x}, \underline{y}) = -\delta^{(D)} \left(\underline{x} - \underline{y} \right),$$

$$\delta^{(D)} \left(\underline{x} - \underline{y} \right) \equiv \delta (x_0 - y_0) \, \delta (x_1 - y_1) \, \delta (x_2 - y_2) \cdot \ldots .$$

Here $\Box(g)$ acts only on the x_μ coordinate, while y_μ here is just a parameter.

---

[1] The point is that $G(l) = G_F(x)$. This is not a coincidence and is a generic phenomenon. The phenomenon is linked to the Wick rotation, which relates the partition function in a statistical mechanical theory (defined in Euclidian space) to the functional integral in a stationary quantum field theory (defined in Minkowski space).

In flat D-dimensional Minkowski space-time this equation is as follows:

$$\left[\partial_{x_0}^2 - \Delta + M^2\right] G(\underline{x}, \underline{y}) \equiv \left[\partial_{x_0}^2 - \partial_{x_1}^2 - \partial_{x_2}^2 - \cdots - \partial_{x_{D-1}}^2 + M^2\right] G(\underline{x}, \underline{y})$$
$$= -\delta^{(D)}\left(\underline{x} - \underline{y}\right).$$

It is invariant under the translations in the space-time, $x_\mu \rightarrow x_\mu + a_\mu$ and $y_\mu \rightarrow y_\mu + a_\mu$, under the Lorentz transformations and rotations. All together that is the Poincaré symmetry of the D-dimensional space-time. As the result of this symmetry any solution of this equation is a function of the geodesic distance $G(\underline{x}, \underline{y}) = G\left(\left|\underline{x} - \underline{y}\right|\right)$.

To solve the equation under consideration we will make the Fourier transformation:

$$G\left(\left|\underline{x} - \underline{y}\right|\right) = \int \frac{d^D p}{(2\pi)^D} \tilde{G}(\underline{p}) e^{-i \underline{p}(\underline{x} - \underline{y})}, \quad \text{and} \quad \delta^{(D)}\left(\underline{x} - \underline{y}\right) = \int \frac{d^D p}{(2\pi)^D} e^{-i \underline{p}(\underline{x} - \underline{y})}.$$

Substituting these expressions into the equation for $G(x, y)$, we obtain that:

$$\left[\underline{p}^2 - M^2\right] \tilde{G}(\underline{p}) = 1.$$

As the result, to find $G(|\underline{x} - \underline{y}|)$ one has to calculate the following integral:

$$G(|\underline{x} - \underline{y}|) = \int \frac{d^D p}{(2\pi)^D} \frac{e^{-i \underline{p}(\underline{x} - \underline{y})}}{\underline{p}^2 - M^2}.$$

Let us calculate this integral in the one of the simplest cases—in the two-dimensional Minkowskian space-times. To do that we have to specify the contour of integration in the complex p_0-plane, where $\underline{p} = (p_0, p_1)$.

As we have explained in the previous subsection the Feynman propagator corresponds to the following situation:

$$G_F(|\underline{x} - \underline{y}|) = -i \iint_{-\infty}^{+\infty} \frac{dp_0 \, dp_1}{(2\pi)^2} \frac{e^{-i p_0 (x-y)_0 + i p_1 (x-y)_1}}{p_0^2 - p_1^2 - M^2 + i 0} = \int_{-\infty}^{+\infty} \frac{dp}{2\pi} \frac{e^{-i \sqrt{p^2 + M^2} |t| + i p s}}{2\sqrt{p^2 + M^2}},$$

where to obtain the second equality we have taken the integral over p_0 using the Cauchy theorem and Jordan's lemma, and have made the following redefinitions of the variables $p_1 \rightarrow p$, $(x - y)_0 \rightarrow t$ and $(x - y)_1 \rightarrow s$.

Using the substitution $p = M \sinh \xi$, $\xi \in (-\infty, +\infty)$, one can rewrite this integral as follows:

$$G_F(|\underline{x} - \underline{y}|) = \frac{1}{4\pi} \int_{-\infty}^{+\infty} d\xi \, e^{-i M (|t| \cosh \xi - s \sinh \xi)}.$$

As we have explained above due to the Poincaré symmetry the propagator can depend only on $|\underline{x} - \underline{y}|^2 \equiv t^2 - s^2$. In fact, consider e.g. $t > 0$ and perform a Lorentz transformation over t and s:

$$t' = t \cosh\alpha + s \sinh\alpha, \quad s' = t \sinh\alpha + s \cosh\alpha,$$

where α is constant, then in the last integral the integration variable will be shifted by α: $\xi \to \xi + \alpha$. I.e. the integral will not change.

As the result, if the interval between $\underline{x}$ and $\underline{y}$ is time-like, $t^2 - s^2 > 0$, we can make its Lorentz rotation such that it will be along only the time direction, i.e. $(x - y)_0 = \sqrt{t^2 - s^2}$ and $(x - y)_1 = 0$.

Note that interval $|\underline{x} - \underline{y}| = \sqrt{t^2 - s^2}$ is not a single valued function on the complex plane of $t^2 - s^2$. Hence, the standard prescription is that there is the cut in this complex plane along the positive real axis and the function under consideration is real on the negative real axis of the $t^2 - s^2$ plane. The reason for this is that imaginary part of the the Feynman propagator G_F is proportional to the retarded Green function, as we have explained in the previous subsection. The latter one should vanish for the space-like intervals, as follows from the physical considerations. Namely from causality.

Thus,

$$G_F(|\underline{x} - \underline{y}|) = \frac{1}{4\pi} \int\limits_{-\infty}^{+\infty} d\xi\, e^{-i M \left[\sqrt{t^2 - s^2} - i\,0\right]\cosh\xi} = -\frac{i}{4} H_0^{(2)}\left[M\sqrt{t^2 - s^2} - i\,0\right], \quad (5.16)$$

where the shift $i0$ in the exponent of the integrand is done for the following reason. For time-like separations we have a cut. Hence, one has to specify on which side of the cut the function should be taken. The $i0$ shift makes this specification. If we shift the argument of the exponential as is shown here the integral remains convergent, while if the shift is done via the inverse sign $-i0$, then the integral will become divergent.

Finally, to make the last step in the last equation we have used the derived above integral representation of the Hankel function of the second kind. The integral under consideration can be reduced to the integral representation of the Hankel function after an obvious change of variables and noticing that contributions coming from some parts of the contour cancell each other.

Likewise, if the interval is space-like, $t^2 - s^2 < 0$, we have that:

$$G_F(|\underline{x} - \underline{y}|) = \frac{1}{4\pi} \int\limits_{-\infty}^{+\infty} d\xi\, e^{-i M \left[\sqrt{s^2 - t^2}\right]\sinh\xi} = \frac{1}{2\pi} K_0\left[M\sqrt{s^2 - t^2}\right],$$

which is real as we have been discussing above. This K_0 is essentially the same function as the above $H_0^{(2)}$, if both of them are considered as complex functions on the cutted complex plane of $t^2 - s^2$. Please note the relations between these functions, which have been derived above.

Finally, note also that the same Makdonald K_0 function solves the two-dimensional Klein–Gordon equation on the Euclidian plane. This is not a coincidence, because for the space-like intervals the Klein–Gordon equation is related to the Euclidian Klein–Gordon equation, as we have mentioned at the end of the previous subsection. In fact, consider the equation

$$\left(\Delta_2 - m^2\right) G\left(\vec{x}, \vec{y}\right) = \delta^{(2)}\left(\vec{x} - \vec{y}\right),$$

It is not hard to see that this equation is invariant with respect to the algebra $ISO(2)$. In fact, $\Delta_2 = h_+ h_- = \hat{a}_1^2 + \hat{a}_2^2$ commutes with $h_\pm$ and a_3. Also $\delta^{(2)}\left(\vec{x} - \vec{y}\right)$ does not change under rotations and translations.

This means that the solution of the equation under consideration should be a function as follows:

$$G\left(\vec{x}, \vec{y}\right) = G\left(|\vec{x} - \vec{y}|\right).$$

Now if we introduce a new notation $m\left(\vec{x} - \vec{y}\right) = \vec{r} = (r, \varphi)$, then the equation under consideration acquires the form:

$$\left(\frac{\partial^2}{\partial r^2} + \frac{1}{r}\frac{\partial}{\partial r} + \frac{1}{r^2}\frac{\partial^2}{\partial \varphi^2} - 1\right) G(r) = \delta^{(2)}\left(\vec{r}\right).$$

Taking into account that $G(r)$ does not depend on φ, the equation reduces to

$$\left(\frac{d^2}{dr^2} + \frac{1}{r}\frac{d}{dr} - 1\right) G(r) = \delta^{(2)}\left(\vec{r}\right),$$

which is nothing but the equation for $K_0(r)$. Whether there is a delta–function on the right hand side of the equation or not depends on how one resolves the peculiarities of this function in the complex plane of its argument.

The derivation of other types of propagators and in other dimensions can be performed in a similar way.

Chapter 6
Legendre Polynomials and Spherical Functions

Abstract This section is recorded by MIPT student Tselousov Nikita. It contains the derivation of various properties of the Legendre polynomials and of various relative functions. It also describes the relation of these functions to the representation theory and quantum mechanics.

Consider the three dimensional Laplace operator: $\Delta_3 = \partial_x^2 + \partial_y^2 + \partial_z^2$. In spherical coordinates $(x, y, z) = (r \sin\theta \cos\varphi, \, r \sin\theta \sin\varphi, \, \cos\theta)$, the metric tensor in the three-dimensional Euclidian space has the following form:

$$dl^2 \equiv g_{\mu\nu} x^\mu \, dx^\nu = dr^2 + r^2 \left(d\theta^2 + \sin^2\theta \, d\varphi^2 \right), \quad \mu, \nu = \overline{1, 2, 3}.$$

Hence, as follows from (5.15) the Laplace operator is:

$$\Delta_3 = \frac{1}{r^2} \partial_r r^2 \partial_r + \frac{1}{r^2 \sin\theta} \partial_\theta \sin\theta \partial_\theta + \frac{1}{r^2 \sin^2\theta} \partial_\varphi^2 \equiv \frac{1}{r^2} \partial_r r^2 \partial_r + \frac{\Delta_{\theta,\varphi}}{r^2},$$

where

$$\Delta_{\theta,\varphi} = \frac{1}{\sin\theta} \partial_\theta \sin\theta \partial_\theta + \frac{1}{\sin^2\theta} \partial_\varphi^2 \tag{6.1}$$

is the Laplace operator on the two-dimensional sphere. We are interested in the eigen-functions of the latter operator:

$$\Delta_{\theta,\varphi} Y(\theta, \varphi) = \lambda Y(\theta, \varphi).$$

The eigen-functions that are regular everywhere on the sphere obey the equation:

$$\Delta_{\theta,\varphi} Y_{lm}(\theta, \varphi) = -l(l+1) Y_{lm}(\theta, \varphi), \tag{6.2}$$

© The Author(s), under exclusive license to Springer Nature Switzerland AG 2019
V. Akhmedova and E. T. Akhmedov, *Selected Special Functions*
for Fundamental Physics, SpringerBriefs in Physics,
https://doi.org/10.1007/978-3-030-35089-5_6

where $l \in \mathbb{N}$, $-l \leqslant m \leqslant l$ and

$$Y_{lm}(\theta, \varphi) = u_{lm}(\theta) \, e^{im\varphi}$$

are the so called spherical harmonics. If one will substitute such an expression into Eq. (6.2), he will find that $u_{lm}(\theta)$ obeys the following equation:

$$\left[\frac{1}{\sin \theta} \frac{d}{d\theta} \sin \theta \frac{d}{d\theta} - \frac{m^2}{\sin^2 \theta} + l(l+1) \right] u_{lm}(\theta) = 0. \qquad (6.3)$$

Furthermore, in the case when $m = 0$ after the change of variables $x = \cos \theta$ we obtain the following equation:

$$\left[\frac{d}{dx} \left(1 - x^2 \right) \frac{d}{dx} + l(l+1) \right] u_l(x) = 0, \qquad (6.4)$$

where $x \in [-1, 1]$. This is the so called Legendre equation. Below we will discuss solutions of the both Eqs. (6.3) and (6.4).

6.1 Generating Function and Integral Representation

We will show below that solutions of the Legendre equation (6.4) can be represented as:

$$u_n(x) \equiv P_n(x) = \frac{1}{2^n n!} \frac{d^n}{dx^n} \left[\left(x^2 - 1 \right)^n \right], \quad n \in \mathbb{N}. \qquad (6.5)$$

These are so called Legendre polynomials. Let us present some of them explicitly for low values of n:

$$P_0(x) = 1, \quad P_1(x) = x, \quad P_2(x) = \frac{1}{2} \left(3x^2 - 1 \right) \quad \text{and} \quad P_3(x) = \frac{1}{2} \left(5x^3 - 3x \right), \dots$$

Consider now the following function:

$$w(x, t) = \frac{1}{\sqrt{1 - 2tx + t^2}},$$

where we choose such a value of the square root at $t = 0$ that it is equal to 1. Consider its series expansion:

$$\frac{1}{\sqrt{1 - 2tx + t^2}} = \sum_{n=0}^{\infty} C_n(x) \, t^n,$$

where $|t| < r$ and r is the smallest of the modules of the roots of the equation $1 - 2tx + t^2 = 0$. Then integrating both sides of the last equation along the contour C that encircles $t = 0$ and is in the region of regularity of $w(x, t)$ and using the Cauchy theorem, we obtain that:

$$C_n(x) = \frac{1}{2\pi i} \oint_C \frac{t^{-n-1}}{\sqrt{1 - 2tx + t^2}} \, dt.$$

If we make the following change of variables $\sqrt{1 - 2tx + t^2} = 1 - tu$, then the last integral transforms into:

$$C_n(x) = \frac{1}{2\pi i} \oint_{C'} \frac{(u^2 - 1)^n}{2^n(u - x)^{n+1}} \, du,$$

where the contour C' encircles $u = x$. Then using again the Cauchy theorem in the form that states

$$f^{(n)}(z) = \frac{n!}{2\pi i} \oint_C \frac{f(\xi)}{(\xi - z)^{n+1}} \, d\xi,$$

we find that

$$C_n(x) = \frac{1}{2^n n!} \left[\frac{d^n (u^2 - 1)^n}{du^n} \right]_{u=x} \equiv P_n(x).$$

Thus, we have shown that $w(x, t)$ is indeed the generating function of the polynomials (6.5):

$$\boxed{\frac{1}{\sqrt{1 - 2tx + t^2}} = \sum_{n=0}^{\infty} P_n(x)t^n.} \tag{6.6}$$

Moreover, we have found that the equation:

$$\boxed{P_n(x) = \frac{1}{2\pi i} \oint_{C'} \frac{(u^2 - 1)^n}{2^n(u - x)^{n+1}} \, du} \tag{6.7}$$

is the integral representation of the polynomials. We did not yet show that these polynomials, $P_n(x)$, solve the Legendre equation.

As the side remark let us stress here that the functions $w(x, t)$ and $P_n(x)$ are related to the multipole decomposition in electrostatics. In fact, consider expansion of the function

$$\frac{1}{\left|\vec{R} - \vec{r}\right|} = \frac{1}{\sqrt{R^2 - 2Rr\cos\theta + r^2}}$$

in powers of r/R when $r \ll R$. Then, we obtain that

$$\frac{1}{\left|\vec{R} - \vec{r}\right|} = \frac{1}{R}\, w\,(\cos\theta,\; r/R),$$

if we define $t = r/R$ and $x = \cos\theta$.

6.2 Recurrence Relations

By the direct substitution it is straightforward to see that $w(x, t)$ obeys the following equation:

$$\left(1 - 2xt + t^2\right)\frac{\partial w}{\partial t} + (t - x)\, w = 0.$$

Substituting into this equation the series expansion (6.6), we find that:

$$\left(1 - 2xt + t^2\right)\sum_{n=0}^{\infty} nP_n(x)t^{n-1} + (t - x)\sum_{n=0}^{\infty} P_n(x)t^n = 0.$$

Equating to zero coefficients of t^n for each n separately, we obtain the first recursion relation:

$$\boxed{(n + 1)\, P_{n+1}(x) - (2n + 1)\, xP_n(x) + nP_{n-1}(x) = 0.} \tag{6.8}$$

Similarly it is easy to see that $w(x, t)$ obeys another equation:

$$\left(1 - 2xt + t^2\right)\frac{\partial w}{\partial x} - tw = 0.$$

Then, substituting again into it the series expansion (6.6), we obtain:

$$\left(1 - 2xt + t^2\right)\sum_{n=0}^{\infty} t^n P_n'(x) - \sum_{n=0}^{\infty} t^{n+1} P_n(x) = 0.$$

From here follows another recurrence relation:

$$\boxed{P_{n+1}'(x) - 2xP_n'(x) + P_{n-1}'(x) - P_n(x) = 0.} \tag{6.9}$$

Differentiating (6.8) and subtracting from the obtained equation and from (6.9) first $P'_{n-1}(x)$ and then $P'_{n+1}(x)$, we obtain that:

$$\boxed{P'_{n+1}(x) - x P'_n(x) = (n+1)\, P_n(x),}$$ (6.10)

and

$$\boxed{x P'_n(x) - P'_{n-1}(x) = n\, P_n(x).}$$ (6.11)

Summing the last two equations, one can find the following relation:

$$P'_{n+1}(x) - P'_{n-1}(x) = (2n+1)\, P_n(x).$$

Finally, changing in (6.10) n for $n-1$ and extracting from (6.11) $P'_{n-1}(x)$, we obtain the relation:

$$\left(1 - x^2\right) P'_n(x) = n\, P_{n-1}(x) - n\, x\, P_n(x).$$

Differentiating again this equation and putting into it $P'_{n-1}(x)$ from (6.11), we find that Legendre polynomials (6.5) obey the Legendre Eq. (6.4) with l exchanged for n.

6.3 Orthogonality

Starting with this subsection we will show that the Legendre polynomials compose the complete and orthonormal basis of functions on the interval $x \in [-1,\,1]$. To start, multiply Legendre equation for $P_m(x)$ by $P_n(x)$ and then subtract from the obtained expression the equation for $P_n(x)$ multiplied by $P_m(x)$. The result is:

$$\left[\frac{d}{dx}\left(1 - x^2\right)\frac{d}{dx}P_m(x)\right]P_n(x) - \left[\frac{d}{dx}\left(1 - x^2\right)\frac{d}{dx}P_n(x)\right]P_m(x) +$$

$$+ \left[m\,(m+1)\ -\ n\,(n+1)\right]P_n(x)P_m(x) = 0$$

or

$$\frac{d}{dx}\left\{(1 - x^2)\left[P'_m(x)P_n(x) - P'_n(x)P_m(x)\right]\right\} + (m-n)(m+n+1)P_m(x)P_n(x) = 0.$$

Integrating this equation over the interval $x \in [-1,\,1]$ and noticing that the integral of the first contribution (which is the total derivative) is zero, we find that:

$$(m - n)\,(m + n + 1) \int_{-1}^{1} P_m(x)\,P_n(x)\,dx = 0.$$

Then, if $m \neq n$, we have that

$$\int_{-1}^{1} P_m(x)\,P_n(x)\,dx = 0. \tag{6.12}$$

Now changing in (6.8) n for $n - 1$, then multiplying it by $(2n + 1)P_n(x)$ and finally subtracting from the obtained expression the Eq. (6.8) itself multiplied by $(2n - 1)P_{n-1}(x)$, one can find the relation:

$$n\,(2n + 1)\,P_n^2(x) + (n - 1)\,(2n + 1)\,P_{n-2}(x)\,P_n(x)$$
$$- (n + 1)\,(2n - 1)\,P_{n-1}(x)\,P_{n+1}(x) - n\,(2n - 1)\,P_{n-1}^2(x) = 0.$$

Integrating this equation over $x \in [-1, 1]$ and using (6.12), we obtain that:

$$\int_{-1}^{1} P_n^2(x)\,dx = \frac{2n - 1}{2n + 1} \int_{-1}^{1} P_{n-1}^2\,dx, \quad n = 2, 3, 4 \ldots .$$

Applying this relation several times to reduce n, we find that:

$$\int_{-1}^{1} P_n^2(x)\,dx = \frac{3}{2n + 1} \int_{-1}^{1} P_1^2(x)\,dx = \frac{2}{2n + 1}.$$

Hence,

$$\int_{-1}^{1} P_n^2(x)\,dx = \frac{2}{2n + 1}.$$

Thus, normalized polynomials $\sqrt{\frac{2n+1}{2}}\,P_n(x)$ compose the orthonormal basis of functions on the interval $x \in [-1, 1]$. Its completeness we will show below.

6.4 Asymptotic Form for the Large Index

To find the asymptotic form of the Legendre polynomials as their index l is taken to infinity we will use the quasiclassical method to solve the corresponding equation. Namely, let us change the variable $t = \ln \tan \frac{\theta}{2}$ in the Legendre equation:

$$\left[\frac{d^2}{dt^2} + \frac{l(l+1)}{\cosh^2 t} \right] P_l(t) = 0.$$

We would like to find the approximate form of the solution of this equation as $l \to \infty$. Let us consider the generic equation of the form:

$$\left[\frac{d^2}{dt^2} - \frac{p^2(t)}{\epsilon^2} \right] \psi(t) = 0, \tag{6.13}$$

where $p(t)$ is some given function and we consider the limit $\epsilon \to 0$.

We look for the solution of this equation in the form $\psi(t) = e^{-S(t)/\epsilon}$, where $S(t) = S_0(t) + \epsilon S_1(t) + \epsilon^2 S_2(t) + \dots$. After substitution of this ansatz into (6.13) and taking the limit $\epsilon \to 0$ it is straightforward to find that:

$$S_0(t) = \int^t dt' \, p(t'), \quad \text{and} \quad S_1(t) = \frac{1}{2} \log p.$$

In the concrete case under consideration

$$p(t) \approx i \frac{1}{\cosh t} = i \sin \theta,$$

and $\frac{1}{\epsilon} = \sqrt{l(l+1)} \approx l + \frac{1}{2}$, as $l \to \infty$, if one keeps the two leading terms. Then the approximate solution of the equation under consideration is as follows:

$$P_l(x) \approx \frac{C}{\sqrt{|p|}} \left[e^{(l+\frac{1}{2}) \int^t dt' p(t') - i \frac{\pi}{4}} + e^{-(l+\frac{1}{2}) \int^t dt' p(t') + i \frac{\pi}{4}} \right], \quad \text{as} \quad l \to \infty.$$

Here C is a constant that is fixed by the normalization. Because in the present case $p(t)$ is pure imaginary one has to do a careful analytical continuation from the real values of $p(t)$ into the pure imaginary ones with the uses of the so called Zwaan method and Stocks lines. That is how one rigorously gets the last answer.

In our case $\int^t dt' p(t') = i\theta$ and, hence,

$$\boxed{P_l(\cos \theta) \approx \frac{2 \cos \left[(l + \frac{1}{2}) \theta - \frac{\pi}{4} \right]}{\sqrt{(2l+1)\pi \sin \theta}}, \quad \text{as} \quad l \to \infty,}$$

which is the asymptotic form in question. The coefficient here is fixed from the normalization condition, which was derived in the previous subsection.

6.5   Completeness

Now we are ready to show the completeness of the basis of the Legendre polynomials. Consider the recurrence relation (6.8). Multiply it by $P_n(y)$ and then subtract the same equation with the exchange of x and y. The result is:

$$(n+1)[P_{n+1}(x)P_n(y) - P_{n+1}(y)P_n(x)] - n[P_n(x)P_{n-1}(y) - P_n(y)P_{n-1}(x)]$$
$$= (2n+1)(x-y)P_n(x)P_n(y).$$

Summing this relation over n from 1 to m and using that $P_0(x) = 1$ and $P_1(x) = x$, we obtain:

$$(x-y)\sum_{n=1}^{m}(2n+1)P_n(x)P_n(y) = (m+1)[P_{m+1}(x)P_m(y) - P_{m+1}(y)P_m(x)] - (x-y).$$

The last relation can be rewritten as:

$$\boxed{\sum_{n=0}^{m}\left(n+\frac{1}{2}\right)P_n(x)P_n(y) = \frac{m+1}{2}\frac{P_{m+1}(x)P_m(y) - P_{m+1}(y)P_m(x)}{x-y}.}$$

Now considering the limit $m \to \infty$ in this expression and using on its RHS the asymptotic form of $P_m(x)$ for the large index, similarly to the case of the Hermite polynomials, we find that:

$$\sum_{n=0}^{+\infty}\left(n+\frac{1}{2}\right)P_n(x)\,P_n(y) = \delta(x-y) \equiv \delta(\cos\theta_1 - \cos\theta_2) = \frac{\delta(\theta_1 - \theta_2)}{|\sin\theta_1|},$$

where $x = \cos\theta_1$ and $y = \cos\theta_2$. The obtained relation establishes the completeness of the basis in question.

6.6   Spherical Harmonics

Let us represent the Legendre equation in the following form

$$(1-x^2)\,u_l'' - 2x\,u_l' + l\,(l+1)\,u_l = 0.$$

Then, differentiate it m times and define $V_{lm} \equiv \frac{d^m}{dx^m}u_l$. As the result V_{lm} obeys the equation as follows:

$$(1 - x^2) V''_{lm} - 2(m+1) x V'_{lm} + (l-m)(l+m+1) V_{lm} = 0.$$

Now define a new function $u_{lm} \equiv (1-x^2)^{\frac{m}{2}} V_{lm}$. This function obeys the equation:

$$(1-x^2)u''_{lm} - 2x u'_{lm} + \left[l(l+1) - \frac{m^2}{1-x^2} \right] u_{lm} = 0,$$

which is the same as (6.3), if $x = \cos\theta$. Thus, its solution can be represented as

$$u_{lm}(\cos\theta) \equiv P_l^m(\cos\theta) \equiv \frac{1}{2^l l!} \sin^m\theta \, \frac{d^{l+m}}{d\cos\theta^{l+m}} \left[\cos^2\theta - 1 \right]^l . \qquad (6.14)$$

These functions are referred to as the associated Legendre polynomials. Similarly to the case of the Legendre polynomials one can show that

$$\int_{-1}^{1} P_l^m(x) P_n^m(x) \, dx = 0$$

if $n \neq l$ and that

$$\int_{-1}^{1} \left[P_l^m(x) \right]^2 dx = \frac{2}{2l+1} \frac{(l+m)!}{(l-m)!}$$

where $|m| \leq l$. It can be seen from (6.14) that $P_l^m(z) = 0$ for $m > l$. In all, one can use

$$Y_{lm}(\theta, \varphi) = \sqrt{\frac{2l+1}{4\pi} \frac{(l-m)!}{(l+m)!}} \, P_l^m(\cos\theta) \, e^{im\varphi}$$

as the orthonormal basis of functions on the sphere—of the functions of θ and φ. The coefficient here follows from the normalization condition.

6.7 Relation to the Representation Theory

Similarly to the Hermite polynomials and the Bessel functions the spherical harmonics are related to a symmetry algebra. Namely, spherical harmonics provide a representation of the $SO(3)$ algebra—the algebra of rotations in three dimensions. Its generators obey the following commutation relations:

$$\left[\hat{L}_m, \hat{L}_n\right] = i\,\varepsilon_{mnk}\,\hat{L}_k, \quad m, n, k = \overline{1,3},$$

where ε_{ijk}—is the absolutely antisymmetric tensor.

The differential operators that provide a representation of this algebra are as follows. The operator:

$$\hat{L}_1 = i\,(x_3\partial_2 - x_2\partial_3)$$

generates rotations around the first, $x = x_1$, axis. The operator

$$\hat{L}_2 = i\,(x_1\partial_3 - x_3\partial_1)$$

generates rotations around the second, $y = x_2$, axis. And finally, the operator

$$\hat{L}_3 = i\,(x_2\partial_1 - x_1\partial_2)$$

generates rotations around the third, $z = x_3$, axis. It is straightforward to show by the direct calculation that these differential operators obeys the above algebra.

One also frequently uses the following operators:

$$\hat{L}_\pm = \hat{L}_1 \pm i\,\hat{L}_2 = e^{\pm i\varphi}\left(\frac{i}{\tan\theta}\partial_\varphi \pm \partial_\theta\right)$$

and $\hat{L}_3 = -i\partial_\varphi$ as the generators of the algebra. Here θ and φ are the coordinates on the sphere. In terms of the latter the commutation relations acquire the following form:

$$\left[\hat{L}_+, \hat{L}_-\right] = \hat{L}_3, \quad \left[\hat{L}_+, \hat{L}_3\right] = -2\,\hat{L}_+, \quad \left[\hat{L}_-, \hat{L}_3\right] = 2\,\hat{L}_-.$$

Furthermore, it is also straightforward to see that the operator:

$$\hat{L}^2 = \hat{L}_1^2 + \hat{L}_2^2 + \hat{L}_3^2 = -\Delta_{\theta,\varphi}$$

commutes with all generators of this algebra:

$$\left[\hat{L}^2, \hat{L}_1\right] = \left[\hat{L}^2, \hat{L}_2\right] = \left[\hat{L}^2, \hat{L}_3\right] = 0.$$

Such an operator is referred to as Casimir one. As we see it coincides with the above defined Laplace operator on the sphere.

One can show that the spherical harmonics obey the following relations:

$$\hat{L}_3 Y_{lm} = -m Y_{lm}, \quad \hat{L}_+ Y_{l\,m-1} = \sqrt{(l+m)(l-m+1)}\, Y_{lm},$$

$$\text{and} \quad \hat{L}_- Y_{lm} = \sqrt{(l+m)(l-m+1)}\, Y_{l\,m-1}.$$

I.e. the vector $\left(Y_{l,-l}, Y_{l,-l+1}, \ldots, Y_{l,l-1}, Y_{l,l}\right)$ for each l separately composes a $(2l + 1)$-dimensional representation of the $SO(3)$ algebra.

6.8 Integral Representation of $P_n^m(\cos\theta)$

A solution of the Laplace equation, $\Delta f(x, y, z) = 0$, in three dimensions can be represented as:

$$f(x, y, z) = (z + ix \cos u + iy \sin u)^n,$$

where $n \in \mathbb{N}$ and $u \in [-\pi, \pi]$. It is straightforward to see that this function indeed solves the Laplace equation as a corollary of the fact that $\cos^2 u + \sin^2 u - 1 = 0$. Furthermore the function

$$V_{lm}(x, y, z) = \int_{-\pi}^{\pi} \left[z + ix \cos u + iy \sin u\right]^l e^{imu}\, du$$

also solves the Laplace equation.

If we make the change of variables to the spherical coordinates $(x, y, z) = (r \sin\theta \cos\varphi, r \sin\theta \sin\varphi, r \cos\theta)$ then, the last function acquires the form:

$$V_{nm}(r, \theta, \varphi) = r^n e^{im\varphi} \int_{-\pi}^{\pi} \left[\cos\theta + i \sin\theta \cos u\right]^n e^{imu}\, du.$$

Substituting this function into the Laplace equation written in the radial coordinates and observing that r^n is the eigen-function of the radial part of the operator, because

$$\partial_r r^2 \partial_r r^n = n(n+1) r^n,$$

and using the relation

$$\partial_\varphi^2 e^{im\varphi} = -m^2 e^{im\varphi},$$

we can see that

$$
u_{lm}(\cos\theta) \equiv \int_{-\pi}^{\pi} [\cos\theta + i \sin\theta \cos u]^l \cos(mu)\, du \tag{6.15}
$$

solves the Legendre equation and is regular for all θ.

Moreover, one can show that actually

$$
P_l^m(\cos\theta) = \frac{i^m\,(l+m)!}{2\pi\,l!}\, u_{lm}(\cos\theta). \tag{6.16}
$$

In fact, consider the integral representation (6.7) of the Legendre polynomials. Let us take as C' the circle of radius $\sqrt{|x^2-1|}$ with the center at $u = x$, i.e $u = x + \sqrt{x^2-1}\,e^{i\varphi}$ on C'. Then, according to (6.7)

$$
P_l(x) = \frac{1}{2\pi}\int_{-\pi}^{\pi} d\varphi \left[\frac{x^2 + 2x\sqrt{x^2-1}\,e^{i\varphi} + (x^2-1)\,e^{i2\varphi} - 1}{2\sqrt{x^2-1}\,e^{i\varphi}}\right]^l
$$

$$
= \frac{1}{\pi}\int_{0}^{\pi}\left[x + i\sqrt{x^2-1}\,\cos\varphi\right]^l d\varphi.
$$

Changing the integration variable as $x = \cos\theta$, we obtain that:

$$
P_l(\cos\theta) = \frac{1}{\pi}\int_{0}^{\pi} [\cos\theta + i\,\sin\theta\,\cos\varphi]^l\, d\varphi \equiv P_l^0(\cos\theta),
$$

which shows the agreement between the two integral representations for the simplest case of the ordinary Legendre polynomials. Using the relation (6.14) between the associated Legendre polynomials and the ordinary ones, from the last integral representation one can show that (6.15) indeed defines $P_l^m(\cos\theta)$.

Finally, using the integral representation (6.15) and the relation (6.16), one can show that

$$
P_l^{-m}(\cos\theta) = (-1)^m \frac{(l-m)!}{(l+m)!} P_l^m(\cos\theta). \tag{6.17}
$$

This relation will be used below.

6.9 Addition or Summation Theorems

Consider some function $f(\theta, \varphi)$ and perform its expansion in spherical harmonics:

$$
\begin{aligned}
f(\theta, \varphi) &= \sum_{n=0}^{\infty} \sum_{m=-n}^{n} A_{nm} P_n^m(\cos\theta) e^{im\varphi} \\
&= \frac{1}{4\pi} \oint d\Omega' f(\theta', \varphi') \sum_{n=0}^{\infty} (2n+1) \sum_{m=-n}^{n} \frac{(n-m)!}{(n+m)!} P_n^m(\cos\theta) P_n^m(\cos\theta') e^{im(\varphi-\varphi')} \\
&= \frac{1}{4\pi} \oint d\Omega' f(\theta', \varphi') \sum_{n=0}^{\infty} (2n+1) \sum_{m=0}^{n} \epsilon_m \frac{(n-m)!}{(n+m)!} P_n^m(\cos\theta) P_n^m(\cos\theta') \cos[m(\varphi-\varphi')],
\end{aligned}
$$

$$(6.18)$$

where

$$
\epsilon_m = \begin{cases} 1, & m = 0 \\ 2, & m > 0, \end{cases}
$$

and we have used the expression for the Fourier coefficients A_{mn} via $f(\theta, \varphi)$ itself. Note that according to (6.17) $P_n^{-m}(t)$ and $P_n^m(t)$ are not linearly independent. Hence, we have to sum above only over the complete basis of $P_n^m(\cos\theta)$ for $n \in (|m|, \infty)$.

Consider the integral kernel in Eq. (6.18):

$$
F(\theta, \varphi \mid \theta', \varphi') \equiv \sum_{n=0}^{\infty} (2n+1) \sum_{m=0}^{n} \epsilon_m \frac{(n-m)!}{(n+m)!} P_n^m(\cos\theta) P_n^m(\cos\theta') \cos[m(\varphi-\varphi')].
$$

$$(6.19)$$

Put in this function $\theta = 0$ and take into account that $P_n(1) = 1$ and $P_n^m(1) = 0$, for $m > 0$. Then we obtain that:

$$
F(\theta = 0, \varphi \mid \theta', \varphi') = \sum_{n=0}^{\infty} (2n+1) P_n(\cos\theta').
$$

$$(6.20)$$

But by an $SO(3)$ rotation we can put $\theta = 0$ to any other point on the sphere, as is shown on the figure:

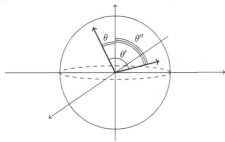

From this figure it can be seen that:

$$\cos \theta' = \cos \theta \cos \theta'' + \sin \theta \sin \theta'' \cos(\varphi - \varphi'') \equiv \left(\vec{X}_1, \vec{X}_2 \right),$$

where $\vec{X}_1$ and $\vec{X}_2$ are two three-dimensional unit vectors, whose ends are sitting on the sphere.

Hence, combining this equation and (6.19) with (6.20) we obtain the relation:

$$P_n \left[\cos \theta \cos \theta' + \sin \theta \sin \theta' \cos(\varphi - \varphi') \right] = \sum_{m=0}^{n} \epsilon_m \frac{(n - m)!}{(n + m)!} P_n^m(\cos \theta) P_n^m(\cos \theta') \cos\left[m(\varphi - \varphi') \right],$$

which also can be written as:

$$P_n \left[\cos \theta \cos \theta' + \sin \theta \sin \theta' \cos(\varphi - \varphi') \right] = \sum_{m=-n}^{n} \frac{(n - m)!}{(n + m)!} P_n^m(\cos \theta) P_n^m(\cos \theta') e^{im(\varphi - \varphi')}.$$

This is the so called summation formula for the Legendre polynomials.

6.10 Legendre Functions $P_\nu^\mu(z)$ for $\mu, \nu \in \mathbb{C}$

One can generalize the Legendre equation to the following one:

$$(1 - z^2) u_\nu'' - 2 z u_\nu' + \nu (\nu + 1) u_\nu = 0, \tag{6.21}$$

where $z \in \mathbb{C}$ and also $\nu \in \mathbb{C}$.

Its solutions are referred to as Legendre functions and are denoted as $P_\nu(z)$ and $Q_\nu(z)$. Also one can define the adjoint Legendre functions as:

$$P_\nu^m(z) = (1 - z^2)^{\frac{m}{2}} \frac{d^m P_\nu(z)}{dz^m} \quad \text{and} \quad Q_\nu^m(z) = (1 - z^2)^{\frac{m}{2}} \frac{d^m Q_\nu(z)}{dz^m}.$$

They solve the following equation:

$$(1 - z^2) u_\nu'' - 2 z u_\nu' + \left[\nu (\nu + 1) - \frac{m^2}{1 - z^2} \right] u_\nu = 0,$$

where $m \geqslant 0$ and can take any integer value. Note that unlike $P_l^m(\cos \theta)$ the functions $P_\nu^m(z)$ and $Q_\nu^m(z)$ are <u>not</u> regular everywhere on the sphere.

Finally, one can define $P_\nu^\mu(z)$ and $Q_\nu^\mu(z)$ with $\mu \in \mathbb{C}$, which solve the equation as follows:

$$(1 - z^2) u_\nu'' - 2 z u_\nu' + \left[\nu (\nu + 1) - \frac{\mu^2}{1 - z^2} \right] u_\nu = 0.$$

Solutions of such an equation we discuss in the next section.

Chapter 7
Hypergeometric Functions

Abstract A part of this section is recorded by MIPT student Kishmar Nikolay. It contains the derivation of various properties of the hypergeometric function. It also contains an elementary description of the relation of this function to the Riemann geometry and to the Green function of the Kleyn–Gordon equation on the sphere.

The so called hypergeometric equation has the following form:

$$z(1 - z)y'' + [c - (a + b + 1)z]\, y' - aby = 0. \tag{7.1}$$

It is the linear second order differential equation with three so called regular peculiarities at $z = 0$, 1, ∞ in the complex z-plane. Thus, its solution is a function on the two dimensional Riemann sphere with three puncture points. To understand the latter moment please recall the stereographic projection between the sphere and the two-dimensional plane.

Let $y' = f(z)\, y$ be a (system of) differential equation(s) on the complex (vector) function, y, of the complex variable z. A peculiarity at $z = z_0$ is referred to as regular if any solution of the system in question behaves in the vicinity of z_0 as a linear combination of the functions of the form $(z - z_0)^\lambda \log^k(z - z_0)$ where $\lambda \in \mathbb{C}$, $k \in \mathbb{N}$.

- E.g. the linear first order differential equation $y' = \lambda \frac{y}{z^k}$ has the regular behavior at $z = 0$ only if $k = 1$. But when $k = 2, 3, \ldots$ the behavior is irregular. In fact, $y = C z^\lambda$ solves this equation for $k = 1$, while when $k > 1$ we have that $y = C \exp\left[-\frac{\lambda}{(k-1)z^{k-1}}\right]$, which is irregular at $z = 0$.

- Similarly the equation $y' = y$ has the only irregular peculiarity at $z = \infty$. In fact, after the change of variables $z = w^{-1}$, we have that $\frac{d}{dz} = -w^2 \frac{d}{dw}$ and $z \to \infty$ corresponds to $w \to 0$. The equation $y' = y$ acquires the form $\dfrac{dy}{dw} = -\dfrac{y}{w^2}$. As the result, the function $y = C e^{\lambda z} = C e^{\frac{\lambda}{w}}$ has the irregularity at $z = \infty$, i.e. at $w = 0$.

- A linear first order differential equation has to have peculiarities on the entire complex z-plane (including infinity). If these peculiarities are regular, then their minimum number is two. E.g. any equation with two peculiar points by a change

© The Author(s), under exclusive license to Springer Nature Switzerland AG 2019
V. Akhmedova and E. T. Akhmedov, *Selected Special Functions*
for Fundamental Physics, SpringerBriefs in Physics,
https://doi.org/10.1007/978-3-030-35089-5_7

of variables can be transformed into the following form $y' = \lambda \frac{y}{z}$, which has regular peculiarities at $z = 0$ and $z = \infty$.

• A system of linear differential equations $y'_i = A_{ij}(z) y_j$ has a regular peculiarity at $z = z_0$, if the matrix function $A_{ij}(z)$ has at z_0 the following form:

$$A_{ij}(z) = \frac{B_{ij}(z)}{z - z_0},$$

where $B_{ij}(z)$ is regular at z_0. In fact, then the equation $\vec{y}' = \hat{A}(z) \vec{y}$ can be rotated to

$$\hat{\Omega} \vec{y}' = \hat{\Omega} \hat{A} \hat{\Omega}^{-1} \hat{\Omega} \vec{y}$$

near $z = z_0$. Choosing $\hat{\Omega}$ such that $\hat{\Omega} \hat{B} \hat{\Omega}^{-1} = Diag(B_i)$ and defining $\hat{\Omega} \vec{y} = \vec{u}$ one can see that in the vicinity of $z = z_0$, the system reduces to

$$u'_i = \frac{B_i}{z - z_0} u_i,$$

whose solution has the following form:

$$u_i = C(z - z_0)^{B_i},$$

in the vicinity of $z = z_0$. Thus, if we take a path around $z = z_0$ in the complex z-plane the vector $\vec{y} = (y_i)$ is transformed by the so called monodromy matrix $\hat{M}$, which can be deduced from the behavior of $\vec{u}$ in the vicinity of $z = z_0$ and its relation to $\vec{y}$. Obviously this matrix does not depend on the form of the path around $z = z_0$. It just depends on the choice of the peculiar point z_0 and on properties of $B_{ij}(z_0)$.

• Linear second order differential equation $y'' + a(z)y' + b(z)y = 0$ has at $z = z_0$ a regular peculiarity, if $a(z)$ has a pole at $z = z_0$ of the order not higher than one, while $b(z)$ has a pole of the order not higher than two. In fact, this differential equation is equivalent to the following system:

$$\begin{cases} y'_1 = \bar{y}_2 \\ \bar{y}'_2 = -a(z)\,\bar{y}_2 - b(z)\,y_1 . \end{cases}$$

If we make a change $y_2 = (z - z_0)\bar{y}_2$, then the system acquires the following form:

$$\begin{cases} y'_1 = \dfrac{y_2}{z - z_0} \\ y'_2 = \left[\dfrac{1}{z - z_0} - a(z) \right] y_2 - (z - z_0)\,b(z)\,y_1 . \end{cases}$$

Thus, the situation is reduced to the previous case.
• Consider now the equation of the following form:

$$y'' + \frac{p(z)}{z - z_0}y' + \frac{q(z)}{(z - z_0)^2}y = 0,$$

where $p(z)$ and $q(z)$ are regular at z_0. Because we have a regular peculiarity at z_0, the solution at the vicinity of this point behaves as:

$$y(z) = (z - z_0)^\lambda U(z),$$

where $U(z)$ is regular at z_0.

Thus, let us look for a solution of the form

$$y(z) = (z - z_0)^\lambda + a_1(z - z_0)^{\lambda+1} + a_2(z - z_0)^{\lambda+2} + \dots . \tag{7.2}$$

Note that the radius of the convergence of the series under consideration is not zero.

Substituting this expression into the differential equation under consideration, and equating to zero the coefficients of the smallest power $(z - z_0)^{\lambda-2}$ here, we obtain that

$$\lambda(\lambda - 1) + p(z_0)\lambda + q(z_0) = 0, \tag{7.3}$$

which is necessary to obey to have such a behavior of $y(z)$ as above. The roots $\lambda_{1,2}$ are referred to as exponents of the peculiarity.

It can be proved that if $\lambda_1 - \lambda_2 \notin \mathbb{Z}$, then each of these λ_1 and λ_2 provides an independent solution of the second order differential equation. Otherwise we encounter the situation which is similar to the one we have met when were defining $Y_m(z)$ Bessel function. In the latter case it is necessary to add logarithmic terms to (7.2) on top of the power like.

7.1 Behavior in the Vicinities of the Peculiar Points

Let us apply the above machinery to the hypergeometric equation. It has peculiarities at $z = 0$, 1 and ∞. In the vicinity of $z = 0$ the equation has the form:

$$y'' + \frac{c}{z}y' - \frac{ab}{z} \approx 0.$$

Thus, $p(z_0) = c$, $q(z_0) = 0$. Hence, Eq. (7.3) reduces to $\lambda(\lambda - 1) + c\lambda = 0$ with the roots $\lambda_1 = 0$ and $\lambda_2 = 1 - c$. Solution corresponding to $\lambda_1 = 0$ will be considered in detail below and is represented by the so called hypergeometric series.

In the vicinity of $z = 1$, we obtain:

$$y'' + \frac{a + b + 1 - c}{z - 1}y' + \frac{ab}{z - 1}y \approx 0.$$

Thus, $p(z_0) = a + b + 1 - c$, while again $q(z_0) = 0$ and (7.3) reduces to $\lambda(\lambda - 1) + (a + b + 1 - c)\lambda = 0$, with the roots $\lambda_1 = 0$ and $\lambda_2 = c - a - b$. Hence, we have one regular at $z = 1$ solution and another one that has the form $y = (z - 1)^{c-a-b} U(z)$, where $U(z)$ is analytic at $z = 1$. (This is all true under the condition that $c - a - b \notin \mathbb{Z}$, as we have mentioned at the end of the previous subsection.)

In the vicinity of $z = \infty$, let us make the change $z \to w = \dfrac{1}{z}$. Hence, using

that $\dfrac{dy}{dz} = -w^2 \dfrac{dy}{dw}$ and $\dfrac{d^2y}{dz^2} = w^4 \dfrac{d^2y}{dw^2} + 2w^3 \dfrac{dy}{dw}$, we find that the hypergeometric equation is transformed into:

$$w^2 (w - 1) \frac{d^2y}{dw^2} + \left[w^2 (2 - c) + (a + b - 1)\, w \right] \frac{dy}{dw} - a b y = 0.$$

Hence, (7.3) reduces to $\lambda (\lambda - 1) - (a + b - 1)\lambda + ab = 0$ with the roots $\lambda_1 = a$, $\lambda_2 = b$, The corresponding solutions are:

$$y_1(z) = z^{-a} + a_1 z^{-a-1} + \dots \quad \text{and} \quad y_2(z) = z^{-b} + b_1 z^{-b-1} + \dots .$$

These series are convergent in a vicinity of $z = \infty$.

7.2   Hypergeometric Series

Let us look for a solution of the hypergeometric equation in the following form:

$$y(z) = z^s \sum_{k=0}^{\infty} C_k z^k,$$

with some constant s.

Substituting this expression into the hypergeometric equation, we find that:

$$\sum_{k=0}^{\infty} C_k z^{s+k-1} (s + k) (s + k - 1 + c) - \sum_{k=0}^{\infty} C_k z^{s+k} (s + k + a) (s + k + b) = 0.$$

Thus, equating to zero coefficients of z^n for each n separately, we find that:

$$C_0\, s\, (s - 1 + c) = 0 \quad \text{and}$$

$$C_k (s + k) (s + k - 1 + c) - C_{k-1} (s + k - 1 + a) (s + k - 1 + b) = 0, \quad k = 1, 2, \dots .$$

The first equation is then solved by either $s = 0$ or by $s = 1 - c$. Let us assume that $c \notin \mathbb{N}$ and choose $s = 0$. Then for C_k we obtain:

$$C_k = \frac{(k - 1 + a)(k - 1 + b)}{k(k - 1 + c)} C_{k-1}, \quad \text{where} \quad k = 1, 2, \ldots .$$

From here, if we choose $C_0 = 1$, it follows that:

$$C_k = \frac{(a)_k (b)_k}{(c)_k k!}, \quad k \in \mathbb{N},$$

where we use the following standard notations:

$$(\lambda)_k \equiv \lambda(\lambda + 1) \ldots (\lambda + k - 1) = \frac{\Gamma(\lambda + k)}{\Gamma(\lambda)}, \quad k = 1, 2, \ldots .$$

Note that $(\lambda)_0 = 1$.

Thus, a particular solution of the hypergeometric equation for $c \neq 0, -1, -2, \ldots$ is as follows:

$$y(z) \equiv {}_2F_1(a, b; c; z) \equiv F(a, b; c; z) = \sum_{k=0}^{\infty} \frac{(a)_k (b)_k}{(c)_k k!} z^k.$$

This series is convergent for $|z| < 1$ and is referred to as the hypergeometric series.

Similarly choosing $s = 1 - c$, we obtain that if $c \neq 2, 3, 4, \ldots$, then:

$$C_k = \frac{(k - c + a)(k - c + b)}{k(k + 1 - c)} C_{k-1}, \quad k = 1, 2, \ldots .$$

From here, if $C_0 = 1$, we find that:

$$C_k = \frac{(1 - c + a)_k (1 - c + b)_k}{k!(2 - c)_k}, \quad k \in \mathbb{N}.$$

Thus, for $c \neq 2, 3, 4, \ldots$ the hypergeometric equation also has another particular solution as follows:

$$y(z) = z^{1-c} \sum_{k=0}^{\infty} \frac{(1 - c + a)_k (1 - c + b)_k}{(2 - c)_k k!} z^k = z^{1-c} {}_2F_1(1 - c + a, 1 - c + b; 2 - c; z).$$

It is defined on the complex z-plane inside the disc $|z| < 1$ and for $|\arg z| < \pi$, which means that there is the cut along the negative real z-axis.

If $c \notin \mathbb{Z}$, then both defined above solutions exist simultaneously and are linearly independent. The other solution in the case when $c \in \mathbb{Z}$ should be found in a similar manner to the above definition of the function $Y_n(z)$.

7.3   Integral Representation and Analytical Continuation

The hypergeometric series is defined only for $|z| < 1$. We will show below that there is a $_2F_1(a, b; c; z)$ function which is defined on the entire complex z-plane with the cut $(1, \infty)$. This function for $|z| < 1$ coincides with the hypergeometric series.

Let us assume that Re $c >$ Re $b > 0$. Then, using the integral representations for Γ- and B- functions, we find that:

$$\frac{(b)_k}{(c)_k} = \frac{\Gamma(c)}{\Gamma(b)\Gamma(c-b)} \int_0^1 t^{b-1+k} (1-t)^{c-b-1}\, dt, \quad \text{where} \ \ k \in \mathbb{N}.$$

Substituting this expression into the hypergeometris series, we find that:

$$F(a, b; c; z) = \frac{\Gamma(c)}{\Gamma(b)\Gamma(c-b)} \sum_{k=0}^{\infty} \frac{(a)_k}{k!} z^k \int_0^1 t^{b-1+k} (1-t)^{c-b-1}\, dt =$$

$$= \frac{\Gamma(c)}{\Gamma(b)\Gamma(c-b)} \int_0^1 dt\, t^{b-1} (1-t)^{c-b-1} \sum_{k=0}^{\infty} \frac{(a)_k\, (z\,t)^k}{k!}.$$

Here

$$\sum_{k=0}^{\infty} \frac{(a)_k}{k!} (tz)^k = (1-tz)^{-a}, \quad 0 \leq t \leq 1, \quad |z| < 1.$$

Thus,

$$\boxed{F(a, b; c; z) = \frac{\Gamma(c)}{\Gamma(b)\Gamma(c-b)} \int_0^1 t^{b-1}(1-t)^{c-b-1}(1-tz)^{-a} dt, \quad \text{Re } c > \text{Re } b > 0, \quad |\arg(1-z)| < \pi.}$$

Now the value of this integral can be analytically continued to the entire cutted z-plane. Note that the restriction $|\arg(1-z)| < \pi$ as usual means that the function in question is defined on the z-plane with the cut, $z \in \mathbb{C}\backslash(1, \infty)$. The reason for the presence of the cut is that $z = 1$ and $z = \infty$ are brunching points of the hypergeometric function, as can be seen from the asymptotic behavior of the solutions of the hypergeometric equation at these peculiar points.

7.4   Contour Barnes Integral Representation

Consider the integral

$$\frac{1}{2\pi i} \int_C \frac{\Gamma(a+s)\,\Gamma(b+s)\,\Gamma(-s)}{\Gamma(c+s)} (-z)^s\, ds,$$

where $|\arg(-z)| < \pi|$. We assume here that it is possible to draw such a contour C that the poles of $\Gamma(a+s)\Gamma(b+s)$ are on the left of it, while the poles of $\Gamma(-s)$ are on the right:

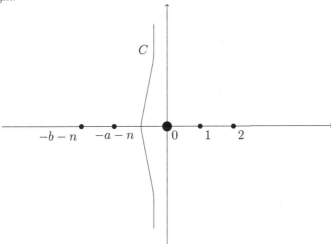

If such a contour is not possible, then $F(a, b; c; z)$ is just a polynomial. Note that $F(-n, b; b; -z) = (1+z)^n$ and also $zF(1, 1; 2, -z) = \log(1+z)$.

It is straightforward to see that the integral under consideration is the analytic function of z in the entire complex z-plane with the cut $|\arg(-z)| < \pi$. Using now the relation (2.2), let us consider an integral as follows:

$$\frac{1}{2\pi i} \int_{C^*} \frac{\Gamma(a+s)\,\Gamma(b+s)\,\pi\,(-z)^s}{\Gamma(c+s)\,\Gamma(1+s)\,\sin(-\pi s)}\,ds,$$

where the contour C^* is defined as shown on the figure:

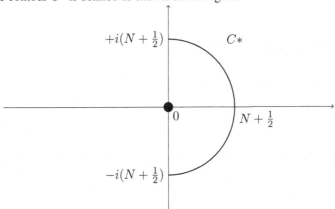

Here N is some large number. It is straightforward to show that if $\log|z| < 0$, i.e. $|z| < 1$, then the integrand in the last expression tends to zero sufficiently fast, i.e.

$$\lim_{N \to \infty} \int_{C*} \frac{\Gamma(a+s)\,\Gamma(b+s)\,\pi\,(-z)^s}{\Gamma(c+s)\,\Gamma(1+s)\,\sin(\pi s)}\,ds = 0$$

Furthermore, the integral of the same expression along the contour as follows:

$$\int_C - \left\{ \int_{-i\infty}^{-(N+\frac{1}{2})i} + \int_{C*} + \int_{(N+\frac{1}{2})i}^{+i\infty} \right\},$$

is equal to $-2\pi i$ times the sum of residues of the integrand at $s = 0, 1, \ldots, N$ according to the Cauchy theorem. Here the contour C is defined above in this subsection.

Let $N \to \infty$ then the integrals inside the curly brackets in the last expression tend to zero, when $|\arg(-z)| < \pi$ and $|z| < 1$. Hence, we have that

$$\frac{1}{2\pi i} \int_C \frac{\Gamma(a+s)\,\Gamma(b+s)\,\Gamma(-s)}{\Gamma(c+s)} (-z)^s\,ds = \lim_{N \to \infty} \sum_{n=0}^{N} \frac{\Gamma(a+n)\,\Gamma(b+n)}{\Gamma(c+n)\,n!}\,z^n,$$

where on the right-hand side of this equation we have the sum over the residues of the integrand under consideration at $s = n$ and $n = \overline{0, N}$. Thus, using the hypergeometric series, we obtain that:

$$\boxed{\frac{\Gamma(a)\,\Gamma(b)}{\Gamma(c)}\,F(a,b;c;z) = \frac{1}{2\pi i} \int_C \frac{\Gamma(a+s)\,\Gamma(b+s)\,\Gamma(-s)}{\Gamma(c+s)} (-z)^s\,ds, \qquad |z| < 1, \quad \arg(-z) < \pi.}$$

This equation provides the so called contour Barnes integral representation of the hypergeometric function.

7.5   Elementary Properties

The hypergeometric series obeys the following relation:

$$\boxed{c\,(c+1)\,F(a,b;c;z) = c\,(c-a+1)\,F(a,b+1;c+2;z) + a\,[c-(c-b)z]\,F(a+1,b+1;c+2;z).}$$

In fact, after the substitution of the hypergeometric series into the RHS of this equation and the collection of all the multipliers of the power z^k for each k separately, we find that:

$$c(c-a+1)\frac{(a)_n(b+1)_k}{(c+2)_k\,k!}+ac\frac{(a+1)_k(b+1)_k}{(c+2)_k\,k!}-a(c-b)\frac{(a+1)_{k-1}(b+1)_{k-1}}{(c+2)_{k-1}\,(k-1)!}=$$

$$=\frac{(a)_k(b)_k}{(c+2)_k\,k!}\left\{c(c-a+1)\frac{b+k}{b}+ac\frac{a+k}{a}\frac{b+k}{b}-a(c-b)\frac{(c+k+1)_k}{ab}\right\}=$$

$$=\frac{(a)_k(b)_k}{(c+2)_k\,k!}(c+k)(c+k+1)=c(c+1)\frac{(a)_k(b)_k}{(c)_k\,k!}.$$

Hence, the equation in question follows.

Also it is easy to see that the hypergeometric series is symmetric under the exchange of a and b, i.e.:

$$\boxed{F(a,b;c;z)=F(b,a;c;z).}$$

Furthermore, differentiating the series, we obtain that:

$$\frac{d}{dz}F(a,b;c;z)=\sum_{k=1}^{\infty}\frac{(a)_k(b)_k}{(c)_k\,(k-1)!}z^{k-1}=\sum_{k=0}^{\infty}\frac{(a)_{k+1}(b)_{k+1}}{(c)_{k+1}\,k!}z^k$$

$$=\frac{ab}{c}\sum_{k=0}^{\infty}\frac{(a+1)_k(b+1)_k}{(c+1)_k\,k!}z^k.$$

Hence,

$$\boxed{\frac{d}{dz}F(a,b;c;z)=\frac{ab}{c}F(a+1,b+1;c+1;z).}$$

Repeating such a differentiation several times, we obtain that:

$$\boxed{\frac{d^m}{dz^m}F(a,b;c;z)=\frac{(a)_m(b)_m}{(c)_m}F(a+m,b+m;c+m;z).}$$

To simplify the equations below, let us define:

$$F(a,b;c;z)\equiv F;$$

$$F(a\pm1,b;c;z)\equiv F(a\pm1);$$

$$F(a,b\pm1;c;z)\equiv F(b\pm1);$$

$$F(a,b;c\pm1;z)\equiv F(c\pm1).$$

Then, we have the following elementary relations:

$$\boxed{(c-a-b)\,F+a(1-z)\,F(a+1)-(c-b)\,F(b-1)=0;}$$

$$(c - a - 1) F + a F(a + 1) - (c - 1) F(c - 1) = 0;$$

$$c (1 - z) F - c F(a - 1) + (c - b) F(c + 1) = 0.$$

To obtain e.g. the first relation we substitute the hypergeometric series into its LHS. Then:

$$(c - a - b) F + a (1 - z) F(a + 1) - (c - b) F(b - 1) =$$

$$= \sum_{k=1}^{\infty} \left[(c - a - b) \frac{(a)_k (b)_k}{(c)_k k!} + a \frac{(a + 1)_k (b)_k}{(c)_k k!} \right.$$

$$\left. -(c - b) \frac{(a)_k (b - 1)_k}{(c)_k k!} - a \frac{(a + 1)(k - 1) (b)_{k-1}}{(c)_{k-1} (k - 1)!} \right] z^k$$

$$= \sum_{k=1}^{\infty} \frac{(a)_k (b)_{k-1}}{(c)_k k!} z^k [(c - a - b) (b + k - 1)$$

$$+ (a + k) (b + k - 1) - (c - b) (b - 1) - (c + k - 1) k] = 0$$

The other two equations can be shown in a similar way. From the already obtained relations and the symmetry $F(a, b; c; z) = F(b, a; c; z)$ one can find also other elementary properties of the hypergeometric function.

7.6   Functional Relations Between Hypergeometric Functions

Let us consider the $SL(2, Z)$ transformation in the complex z-plane as follows:

$$z' = \frac{az + b}{cz + d},$$

where the matrix:

$$\begin{pmatrix} a & b \\ c & d \end{pmatrix}$$

belongs to the so called $SL(2, Z)$ group. This is the group of 2×2 matrices, whose elements obey the conditions as follows: $a, b, c, d \in Z$ and $a d - b c = 1$.

These transformations exchange the points $z = 0, 1$ and ∞ between each other. In fact, apart from the trivial transformation $z = z'$, they contain the following ones:

$$z' = \frac{z}{z - 1}, \quad z' = 1 - z, \quad z' = \frac{1}{1 - z}, \quad z' = \frac{1}{z} \text{ and } z' = \frac{z - 1}{z}.$$

Let $z \in \mathbb{C}\backslash(1, \infty)$, which is another notation of the complex z-plane with the cut from 1 to infinity along the real axis, and assume that Re $c >$ Re $b > 0$. Using the integral representation for the hypergeometric function and changing in it $s = 1 - t$, we find that:

$$F(a, b; c; z) = \frac{\Gamma(c)}{\Gamma(b)\Gamma(c-b)} \int_0^1 ds \, s^{c-b-1} (1-s)^{b-1} (1-z+sz)^{-a} =$$

$$= (1-z)^{-a} \frac{\Gamma(c)}{\Gamma(b')\Gamma(c-b')} \int_0^1 ds \, s^{b'-1} (1-s)^{c-b'-1} (1-sz')^{-a},$$

where $b' = c - b$, and $z' = \dfrac{z}{z-1}$. Here also $z' \in \mathbb{C}\backslash(1, \infty)$ and Re $c >$ Re $b' > 0$.
Thus, we obtain that:

$$\boxed{F(a, b; c; z) = (1-z)^{-a} F\left(a, c-b; c; \frac{z}{z-1}\right), \quad \text{where} \quad |\arg(1-z)| < \pi.}$$

$$(7.4)$$

This relation can be analytically continued beyond the region Re $c >$ Re $b' > 0$.
   Another relation can be obtained from the above one and the symmetry $F(a, b; c; z) = F(b, a; c; z)$:

$$\boxed{F(a, b; c; z) = (1-z)^{-b} F\left(c-a, b; c; \frac{z}{z-1}\right), \quad \text{where} \quad |\arg(1-z)| < \pi.}$$

$$(7.5)$$

Using these two relations one after another, we find that:

$$F(a, b; c; z) = (1-z)^{-a} \left(1 - \frac{z}{z-1}\right)^{-(c-b)} F(c-a, c-b; c; z)$$

or

$$\boxed{F(a, b; c; z) = (1-z)^{c-a-b} F(c-a, c-b; c; z), \quad \text{where} \quad |\arg(1-z)| < \pi.}$$

Furthermore, as we have pointed out above a generic solution of the hypergeometric equation can be represented as:

$$y(z) = A_1 F(a, b; c; z) + A_2 z^{1-c} F(1-c+a, 1-c+b; 2-c; z)$$

for $c \notin \mathbb{Z}$. This can be done on the complex plane with two cuts $|\arg(1-z)| < \pi$ and $|\arg z| < \pi$. Here $A_{1,2}$ are some constants.

Under the change $z' = 1 - z$ the above region in the complex z-pane is transformed into the one, which also contains two cuts $\left| \text{agr} \left(1 - z' \right) \right| < \pi$ and $\left| \arg z' \right| < \pi$ and the hypergeometric equation is transformed into another equation of the same type, but with different parameters: $a' = a$, $b' = b$ and $c' = 1 + a + b - c$. Hence, the linear combination of the two functions:

$$y(z) = B_1 F(a, b; 1 + a + b - c; 1 - z)$$
$$+ B_2 (1 - z)^{c-a-b} F(c - a.c - b; 1 - a - b + c; 1 - z)$$

also solves the same hypergeometric equation. Here we assume that $a + b - c \notin \mathbb{Z}$, $B_{1,2}$ are some constants and $|\text{agr} (1 - z)| < \pi$, $|\arg z| < \pi$. Thus, there should be a linear relation:

$$F(a, b; c; z) = C_1 F(a, b; 1 + a + b - c; 1 - z)$$
$$+ C_2 (1 - z)^{c-a-b} F(c - a, c - b; 1 - a - b + c; 1 - z),$$

for $a + b - c \notin \mathbb{Z}$ and some constants $C_{1,2}$. This means that we can reexpand one of the vectors of the first basis of solutions of the hypergeometric equation in terms of the second basis of functions solving the same equation. Taking in this relation the limits $z \to +0$ and then $z \to 1 - 0$, we find that:

$$C_1 = \frac{\Gamma(c) \, \Gamma(c - a - b)}{\Gamma(c - a) \, \Gamma(c - b)}$$

and

$$1 = C_1 \frac{\Gamma(1 + a + b_c) \, \Gamma(1 - c)}{\Gamma(1 + a - c) \, \Gamma(1 + b - c)} + C_2 \frac{\Gamma(1 - a - b + c) \, \Gamma(1 - c)}{\Gamma(1 - a) \, \Gamma(1 - b)}.$$

From where one can deduce that:

$$C_2 = \frac{\Gamma(c) \, \Gamma(a + b - c)}{\Gamma(a) \, \Gamma(b)}.$$

To obtain these relations we have taken the limit:

$$\lim_{z \to 1-0} F(a, b; c; z) = \frac{\Gamma(c)}{\Gamma(b) \, \Gamma(c - b)} \int_0^1 t^{b-1} (1 - t)^{c-a-b-1} \, dt = \frac{\Gamma(c) \, \Gamma(c - a - b)}{\Gamma(c - a) \, \Gamma(c - b)}.$$

Thus, the functional relation under consideration is as follows:

$$\boxed{F(a, b; c; z) = \frac{\Gamma(c) \, \Gamma(c - a - b)}{\Gamma(c - a) \, \Gamma(c - b)} F(a, b; 1 + a + b - c; 1 - z) +}$$

$$\boxed{+ \frac{\Gamma(c) \, \Gamma(a + b - c)}{\Gamma(a) \, \Gamma(b)} (1 - z)^{c-a-b} F(c - a, c - b; 1 - a - b + c; 1 - z),} \quad (7.6)$$

where as usual $a + b - c \notin \mathbb{Z}$.

Furthermore, consecutive application of (7.4) and (7.6) leads to the relation as follows:

$$F(a,\, b;\, c;\, z) = (1-z)^{-a}\, \frac{\Gamma(c)\,\Gamma(b-a)}{\Gamma(c-a)\,\Gamma(b)}\, F\left(a,\, c-b;\, 1+a-b;\, \frac{1}{1-z}\right) +$$

$$+(1-z)^{-b}\, \frac{\Gamma(c)\,\Gamma(a-b)}{\Gamma(c-b)\,\Gamma(a)}\, F\left(c-a,\, b;\, 1-a+b;\, \frac{1}{1-z}\right),\quad a-b\notin \mathbb{Z},\quad (7.7)$$

on the complex plane with the cuts $|\arg(-z)| < \pi$ and $|\arg(1-z)| < \pi$. And finally, combining (7.7) and (7.4), (7.5) we obtain the relation

$$F(a,\, b;\, c;\, z) = (-z)^{-a}\, \frac{\Gamma(c)\,\Gamma(b-a)}{\Gamma(c-a)\,\Gamma(b)}\, F\left(a,\, 1+a-c;\, 1+a-b;\, \frac{1}{z}\right) +$$

$$+(-z)^{-b}\, \frac{\Gamma(c)\,\Gamma(a-b)}{\Gamma(c-b)\,\Gamma(a)}\, F\left(b,\, 1+b-c;\, 1+b-a;\, \frac{1}{z}\right),\quad a-b\notin \mathbb{Z},\quad (7.8)$$

on the complex plane with the same cuts as in the previous case. We will use this relation in the next subsection.

All these observations are consequences to the fact that the hypergeometric functions are related to the representations of the $SL(2, Z)$ group and its subgroups via the aforementioned monodromy matrices around the peculiar points $z = 0, 1, \infty$.

7.7 Asymptotic Form for the Large Argument

We have already discussed above the asymptotic behavior of solutions of the hypergeometric equation as $z \to \infty$. Let us consider it again from another perspective. Namely, as $z \to \infty$ the hypergeometric equation reduces to

$$z^2\, y'' + (a+b+1)\, z\, y' + a\, b\, y \approx 0,$$

which is the homogeneous in z equation. Hence, its solution has to have the following form $y(z) \propto z^{\alpha}$. After the substitution of this expression into the equation in question, we obtain that α should solve the equation:

$$\alpha\,(\alpha - 1) + (a+b+1)\,\alpha + a\,b = 0.$$

It has two solutions: $\alpha = -a$ and $\alpha = -b$. Hence, a generic solution of the hypergeometric equation should behave as

$$y(z) \approx C_1\, z^{-a} + C_2\, z^{-b},\quad \text{as}\ \ z \to \infty,$$

which agrees with the previous considerations. Here the complex coefficients $C_{1,2}$ depend on the concrete choice of the hypergeometric function.

For example, for the case of the standard hypergeometric series one can find these constants from the Eq. (7.8). In fact, taking $z \to \infty$ in this equation and using that $F(a, b; c; 0) = 1$ for any a, b and c, with obvious restrictions, we obtain the relation

$$\lim_{z \to \infty} F(a, b; c; z) = (-z)^{-a} \frac{\Gamma(c) \Gamma(b-a)}{\Gamma(c-a) \Gamma(b)} + (-z)^{-b} \frac{\Gamma(c) \Gamma(a-b)}{\Gamma(c-b) \Gamma(a)}, \quad (7.9)$$

which provides the concrete expressions for $C_{1,2}$ constants for the concrete solution.

7.8 Relation to the Legendre Functions

Consider the Legendre equation (6.21) and make the following change of variables in it: $t = (1 - z)/2$. Such a transformation converts (6.21) into:

$$t(1-t) \frac{d^2 u}{dt^2} + (1 - 2t) \frac{du}{dt} + \nu(\nu+1) u = 0,$$

which is the hypergeometric equation with the concrete values of the parameters $a = -\nu, b = \nu + 1$ and $c = 1$.

At the same time, the substitution $t = z^{-2}$ and $u = z^{-\nu-1} v$ converts (6.21) into:

$$t(1-t) \frac{d^2 v}{dt^2} + \left[\left(\nu + \frac{3}{2} \right) - \left(\nu + \frac{5}{2} \right) t \right] \frac{dv}{dt} - \left(\frac{\nu}{2} + 1 \right) \left(\frac{\nu}{2} + \frac{1}{2} \right) v = 0,$$

which is again the special case of the hypergeometric equation corresponding to $a = \frac{\nu}{2} + 1, b = \frac{\nu}{2} + \frac{1}{2}$ and $c = \nu + \frac{3}{2}$.

As the result, the two particular solutions of the Legendre equation can be represented as:

$$u_1(z) = F\left(-\nu, \nu + 1; 1; \frac{1-z}{2} \right), \quad |z - 1| < 2,$$

and

$$u_2(z) = \frac{1}{(2z)^{\nu+1}} F\left(\frac{\nu}{2} + 1, \frac{\nu}{2} + \frac{1}{2}; \nu + \frac{3}{2}; \frac{1}{z^2} \right),$$

where $|z| > 1, |\arg z| < \pi$ and $\nu \neq -1, -2, \ldots$.

It can be shown that the first solution coincides with $P_\nu(z)$—the Legendre function of the first kind:

$$\boxed{P_\nu(z) = F\left(-\nu, \nu + 1; 1; \frac{1-z}{2} \right), \quad |z - 1| < 2,}$$

while the second one is proportional to Q_ν—the Legendre function of the second kind:

$$Q_\nu(z) = \frac{\sqrt{\pi}\,\Gamma(\nu+1)}{\Gamma\left(\nu+\frac{3}{2}\right)(2z)^{\nu+1}} F\left(\frac{\nu}{2}+1, \frac{\nu}{2}+\frac{1}{2}; \nu+\frac{3}{2}; \frac{1}{z^2}\right).$$

where again $|z| > 1$, $|\arg z| < \pi$ and $\nu \neq -1, -2, \ldots$, but one can analytically continue these relations to the entire complex z-plane with the cut.

Using the above relations we can also find that:

$$P_\nu^m(z) = \left(1-z^2\right)^{\frac{m}{2}} \frac{d^m}{dz^m} F\left(-\nu, \nu+1; 1; \frac{1-z}{2}\right) =$$

$$= \frac{\left(1-z^2\right)^{\frac{m}{2}}}{2^m} \frac{(-1)^m}{(1)_m} \frac{(-\nu)_m (\nu+1)_m}{(1)_m} F\left(m-\nu, \nu+m+1; m+1; \frac{1-z}{2}\right)$$

or

$$P_\nu^m(z) = \frac{\Gamma(\nu+m+1)\left(1-z^2\right)^{\frac{m}{2}}}{2^m\,\Gamma(m+1)\,\Gamma(\nu-m+1)} F\left(m-\nu, \nu+m+1; m+1; \frac{1-z}{2}\right),$$

where $|\arg(z-1)| < \pi$, $m \in \mathbb{N}$ and ν is arbitrary. These equations establish the relations between the associated Legendre functions and the hypergeometric ones.

7.9   Application: The Feynman Propagator on the Sphere

Let us discuss an application of the hypergeometric functions to physics. From Eq. (5.15) and from Eq. (6.1) one can find that the Klein–Gordon equation on the two dimensional sphere of unit radius is as follows:

$$\left[\frac{1}{\sin\theta} \partial_\theta \sin\theta\partial_\theta + \frac{1}{\sin^2\theta} \partial_\varphi^2 - M^2\right] G\left(\theta, \varphi|\theta', \varphi'\right) = \frac{\delta\left(\theta-\theta'\right)}{\sin\theta} \delta\left(\varphi-\varphi'\right).$$
$$(7.10)$$

Here $\sin\theta$ in the denominator on the RHS appears as the square root of the determinant of the metric on the sphere:

$$\frac{\delta\left(\theta-\theta'\right)}{\sin\theta} \delta\left(\varphi-\varphi'\right) = \delta\left(\cos\theta-\cos\theta'\right) \delta\left(\varphi-\varphi'\right) \equiv \delta^{(2)}\left(\theta, \varphi|\theta', \varphi'\right).$$

We have used these relations in the section on Legendre polynomials.

The equation for the Green function is invariant under the $SO(3)$ rotations of the sphere. Hence, by such rotations one can put the source point (θ', φ') to the northern pole of the sphere. As the result, $G\,(\theta, \varphi|0, 0)$ becomes independent of φ. In fact, as the consequence of the invariance under rotations the function $G\left(\theta, \varphi|\theta', \varphi'\right)$ should depend only on the geodesic distance between (θ, φ) and $\left(\theta', \varphi'\right)$, which, when $\left(\theta', \varphi'\right)$ coincides with the north pole of the sphere, is equal to θ measured in radians and multiplied by the radius of the sphere. Namely,

$$G\,(\theta, \varphi|0, 0) = g(\cos \theta).$$

we choose here $\cos \theta$ instead of θ itself as the argument of the function just because it is more convenient for the equations that follow. We have been using similar reasoning when were deriving the summation equations for the spherical harmonics.

As the result, as follows from (7.10), when we rotate the coordinate system on the sphere such that $\left(\theta', \varphi'\right)$ coincides with the northern pole and, hence, $z = \cos \theta$, the equation for the Green function is reduced to:

$$\left[\partial_z \left(1 - z^2\right) \partial_z - M^2\right] g(z) = \left[(1 - z^2)\partial_z^2 - 2\,z\,\partial_z - M^2\right] g(z) = \delta^{(2)}\,(\theta, \varphi|0, 0)\,.$$

The equation under consideration has three regular peculiarities at $z = \pm 1$ and $z = \infty$ in the complex z-plane. By the change $z \to (1 \pm z)/2$ one can put the peculiar points to their standard positions—into 0, 1 and ∞. As the result, the equation under consideration acquires the hypergeometric form and its arbitrary solution can be written as:

$$g(z) = A_1\, F\left(h_+, h_-;\, \frac{1}{2};\, \frac{1-z}{2}\right) + A_2\, F\left(h_+, h_-;\, \frac{1}{2};\, \frac{1+z}{2}\right),$$

$$\text{where} \quad h_\pm = \frac{1}{2} \pm \sqrt{\left(\frac{1}{2}\right)^2 - M^2}, \quad (7.11)$$

and the constants $A_{1,2}$ are yet to be fixed by some physical conditions that we will discuss now.

The hypergeometric function multiplying A_1 has the brunching point at $z = 1$, while the one multiplying A_2—at $z = -1$. At the same time, while $z = 1$ corresponds to the situation when $(\theta, \varphi) = \left(\theta', \varphi'\right)$, the case $z = -1$ corresponds to the situation when (θ, φ) coincides with the antipodal point of $\left(\theta', \varphi'\right)$—with the point which sits on the opposite end of the diameter emanating from $\left(\theta', \varphi'\right)$. Thus, in first place on the physical grounds one should choose $A_2 = 0$ in (7.11), because one does not expect any peculiarities of the Green function at the antipodal point of its source. One expects peculiar behavior of the propagator only at the coincidence limit of its two arguments.

To fix the coefficient A_1 let us consider the limit $z \to 1$. As we have just pointed out this corresponds to the case when (θ, φ) is very close to $\left(\theta', \varphi'\right)$. In a small vicinity of its any point the two-dimensional sphere looks just like the flat two-

dimensional Euclidian space. It means that (7.11) should look the same as (5.16) or its K_0 counterpart, when $z \to 1$ and $t^2 - s^2 \to 0$. Similarly to the Y_m case.
    In fact, as $l = \sqrt{t^2 - s^2} \to 0$ the Green function (5.16) behaves as:

$$G_F(l) \approx \frac{1}{2\pi} \log(l - i0), \quad \text{when} \quad l \to 0.$$

This expression can be found both from the $H_0^{(2)}$ and K_0 expressions for the propagator on the two-dimensional plane. At the same time the case of (7.11) corresponds exactly to the situation when $c - a - b = 0$ and all the above given formulas in this section are not applicable for the limit when the argument is taken to zero. In this case we have to have logarithmic terms on top of the power like.
    I fact, if $A_2 = 0$ the limit of (7.11) when $z = \cos l \to 1$ is as follows:

$$g(z) \approx A_1 \log(1 - z).$$

Hence, the Feynman propagator on the sphere has the following form:

$$g(z) = \frac{1}{4\pi} F\left(h_+, h_-; \frac{1}{2}; \frac{1-z}{2} - i0\right),$$

and the constant A_1 is fixed from the relation between the two limiting expressions.

Chapter 8
Degenerate Hypergeometric Function

Abstract This section is recorded by MIPT student Aleksandr Artemev. It contains the derivation of various properties of the degenerate hypergeometric function. It also contains the derivation of the relations of this function to other special functions.

Consider the following limit of the hypergeometric function:

$$\lim_{b \to \infty} F\left(a, b, c; \frac{z}{b}\right) = \lim_{b \to \infty} \sum_{k=0}^{\infty} \frac{(a)_k \, (b)_k}{(c)_k \, k!} \left(\frac{z}{b}\right)^k.$$

In this limit we obtain the following series:

$$F(a, c; z) = \sum_{k=0}^{\infty} \frac{(a)_k}{(c)_k} \frac{z^k}{k!}, \tag{8.1}$$

for $c \neq 0, -1, -2, \ldots$ This is the so called degenerate or confluent hypergeometric function. Note that the series for it converges for $|z| < \infty$.

8.1 Differential Equation

Let us show that the degenerate hypergeometric function obeys the following differential equation

$$z \, y'' + (c - z) \, y' - a \, y = 0, \tag{8.2}$$

for $c \neq 0, -1, -2, \ldots$. In fact, substituting there the series (8.1), we obtain on the LHS of this equation

V. Akhmedova and E. T. Akhmedov, *Selected Special Functions*
for Fundamental Physics, SpringerBriefs in Physics,
https://doi.org/10.1007/978-3-030-35089-5_8

$$\sum_{k=2}^{\infty} \frac{k(k-1)(a)_k}{(c)_k \, k!} z^{k-1} + (c-z) \sum_{k=1}^{\infty} \frac{k(a)_k}{(c)_k \, k!} z^{k-1} - a \sum_{k=0}^{\infty} \frac{(a)_k}{(c)_k \, k!} z^{k}$$

$$= c \frac{(a)_1}{(c)_1} - a + \sum_{k=1}^{\infty} \frac{z^k \, (a)_k}{k! \, (c)_k} \left(k \frac{a+k}{c+k} + c \frac{a+k}{c+k} - k - a \right) = 0.$$

Thus, the equation under discussion is satisfied.

To find the second solution of this second order differential equation, let us substitute there $u = z^{1-c} v$. Then, it transforms into:

$$z \, v'' + (\overline{c} - z) \, v' - \overline{a} \, v = 0,$$

where $\overline{a} = 1 + a - c$, $\overline{c} = 2 - c$. Hence, if $\overline{c} \neq 2, 3, \ldots$ the function

$$\boxed{u = z^{1-c} \, F(1 + a - c, 2 - c; z)}$$

also solves the degenerate hypergeometric equation. Thus, if $c \notin \mathbb{Z}$, then

$$u = A F(a, c; z) + B z^{1-c} \, F(1 + a - c, 2 - c; z)$$

is a general solution for some constants A and B. When $c = -1, -2, \ldots$ one can define G—the second kind degenerate hypergeometric function. Again the situation is similar to the one which we encountered when have been defining the Y_m Bessel function.

Finally, let us point out that the confluent hypergeometric function is defined on the two-dimensional Riemann sphere with two punctures: Two punctures among the three ones of the hypergeometric function are merged in the limit, in which the degenerate hypergeometric function follows from the standard one.

8.2 Integral Representation

Using that

$$\frac{(a)_k}{(c)_k} = \frac{\Gamma(c)}{\Gamma(a) \, \Gamma(c-a)} \cdot \int_0^1 t^{a-1+k} (1-t)^{c-a-1} \, dt,$$

where $k \in \mathbb{N}$, and $\mathrm{Re}\, c > \mathrm{Re}\, a > 0$, we can obtain that:

$$F(a, c; z) = \frac{\Gamma(c)}{\Gamma(a) \, \Gamma(c-a)} \cdot \sum_{k=0}^{\infty} \frac{z^k}{k!} \int_0^1 t^{a-1+k} (1-t)^{c-a-1} \, dt$$

$$= \frac{\Gamma(c)}{\Gamma(a) \, \Gamma(c-a)} \cdot \int_0^1 dt \, t^{a-1} (1-t)^{c-a-1} \sum_{k=0}^{\infty} \frac{(tz)^k}{k!}.$$

As the result,

$$F\left(a,c;z\right) = \frac{\Gamma(c)}{\Gamma(a)\,\Gamma(c-a)} \cdot \int_0^1 dt\, t^{a-1}\,(1-t)^{c-a-1} e^{zt}, \quad \mathrm{Re}\,c > \mathrm{Re}\,a > 0.$$

$$(8.3)$$

Let us make the substitution $t = 1 - s$ in this integral. Then:

$$F\left(a,c;z\right) = \frac{\Gamma(c)\,e^z}{\Gamma(a)\,\Gamma(c-a)} \cdot \int_0^1 ds\,(1-s)^{a-1}\,s^{c-a-1}\,e^{-zs},$$

because $\mathrm{Re}\,c > \mathrm{Re}\,(c-a)$, it follows from here that:

$$F\left(a,c;z\right) = e^z\,F\left(c-a,c;-z\right).$$

Via the analytic continuation one can see that this relation is valid for any values of a and c, except $c = 0, -1, -2, \ldots$

8.3 Laplace Transformation

Consider a differential equation of the form

$$\sum_{m=0}^{N} (a_m + b_m z)\,\frac{d^m y}{dz^m} = 0.$$

Let us look for a solution of this equation in the following form

$$y(z) = \int_C dt\, Z(t)\, e^{zt}, \qquad (8.4)$$

with some contour C and some $Z(t)$ to be specified in a moment.

Taking into account that $z\,e^{zt} = \dfrac{d}{dt}e^{zt}$ and substituting (8.4) into the differential equation, we obtain after the partial integration that the equation is satisfied if

$$\frac{d}{dt}\,(QZ) = PZ,$$

where $P(t) = \sum\limits_{m=0}^{N} a_m t^m$ and $Q(t) = \sum\limits_{m=0}^{N} b_m t^m$, and if boundary contributions at the ends of the contour C are vanishing.

From the last equation for $Z(t)$, we find that

$$Z(t) = \frac{1}{Q(t)} \exp\left(\int^{t} dt' \, \frac{P(t')}{Q(t')} \right).$$

The choice of the contour C in (8.4) is dictated by the fact that the expression ZQe^{zt} should vanish at its ends. Then the natural choice is that it comes along one of the directions from infinity, where ZQe^{zt} is vanishing, and then returns back to infinity along other direction, in which ZQe^{zt} also tends to zero. The number of such directions grows with N which provides the correct number of independent solutions of the above defined N-th order differential equation.

8.4   Another Integral Representation from the Laplace Transformation

Applying the Laplace method to the degenerate hypergeometric equation (8.2), we find that the P and Q polynomials are equal to $P(t) = ct - a$ and $Q(t) = t(t - 1)$. Hence, it is not hard to find that

$$Z(t) = t^{a-1} \, (t - 1)^{c-a-1}.$$

Thus, the solution of the differential equation under consideration is as follows:

$$u_1(z) = \int\limits_{C} dt \, e^{tz} \, t^{a-1} (t - 1)^{c-a-1}.$$

Using the same procedure, for another solution $u = z^{1-c} v$ we obtain

$$u_2(z) = z^{1-c} \int\limits_{C} dt \, e^{tz} \, t^{a-c} (t - 1)^{-a}.$$

Changing the variables $tz \to t$, we find that:

$$u_2(z) = \int\limits_{C} dt \, e^{t} \, t^{a-c} (t - z)^{-a}.$$

It is natural to choose the contour C as shown on the figure:

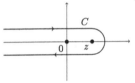

It encircles the peculiarities of the integrand at $t = 0$ and $t = z$ and the product $Z\,Q\,e^{zt}$ vanishes at the ends of this contour. The integral does not have peculiarity at $z = 0$, if C is encircling the above two peculiarities.

We assume that the cuts of t^{a-c} and $(t-z)^{-a}$ are going to $-\infty$ along the real axis and the values of these functions for positive variables are also positive. Then, the second integral under consideration coinsides with $F(a, c; z)$ up to a constant multiplier. To find the latter, let us put $z = 0$. Then,

$$\int_C t^c\, e^t\, dt = \frac{2\pi i}{\Gamma(c)},$$

as follows from the integral representation of Γ-function. Recalling that $F(a, c; 0) = 1$, we find the following expression

$$F(a, c; z) = \frac{\Gamma(c)}{2\pi i} \int_C dt\, e^t\, t^{a-c}\, (t-z)^{-a}. \tag{8.5}$$

For $c = -1, -2, \dots$ the function $F(a, c; z)$ is not defined, which reveals itself via the presence of $\Gamma(c)$. At the same time, the contour integral multiplier itself in (8.5) is well defined even for $c = -1, -2, \dots$.

8.5 Simplest Relations

Changing $t \to t + z$ in the integral representation, we find the relation from the subsection (8.2). Differentiating the integral (8.3) over z and integrating by parts inside the contour allows to find the relations as follows:

$$\frac{d}{dz} F(a, c; z) = \frac{a}{c} F(a+1, c+1; z);$$

$$\frac{z}{c} F(a+1, c+1; z) = F(a+1, c; z) - F(a, c; z);$$

$$a F(a+1, c+1; z) = (a-c) F(a, c+1; z) + c F(a, c; z).$$

These are the simplest relations that the degenerate hypergeometric function obeys.

8.6   Asymptotic Behavior for the Large Argument

For Re $z \to \infty$, the main contribution to the integral (8.5) for $F(a, c; z)$ comes
from the vicinity of $t = z$. Changing the variables $t = z + \xi$ and neglecting the
ξ-dependence in t^{a-c}, we find that

$$F(a, c; z) \approx \frac{\Gamma(c)}{2\pi i} e^z z^{a-c} \int_{C'} d\xi \, e^\xi \, \xi^{-a},$$

where the contour C' is shown on the figure:

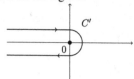

Then,

$$\boxed{F(a, c; z) \approx \frac{\Gamma(c)}{\Gamma(a)} e^z z^{a-c},}$$

as Re $z \to \infty$.

8.7   Relations to Other Functions

Many elementary functions can be expressed via $F(a, c; z)$. E.g.,

$$F(a, a; z) = \sum_{k=0}^{\infty} \frac{z^k}{k!} = e^z.$$

Furthermore,

$$\boxed{F(-n, c, z) = P_n(z),}$$

where $P_n(z)$ is a polynomial in z. For example, the Hermite polynomials follow as:

$$H_{2n}(z) = \sum_{k=0}^{n} (-1)^k \frac{(2n)!}{k!(2n-2k)!} (2z)^{2n-2k} = (-1)^n (2n)! \sum_{k=0}^{n} \frac{(-1)^k (2z)^{2k}}{(n-k)!(2k)!} =$$

$$= (-1)^n \frac{(2n)!}{n!} \sum_{k=0}^{n} \frac{(-n)_k (2z)^{2k}}{(2k)!} = (-1)^n \frac{(2n)!}{n!} \sum_{k=0}^{n} \frac{(-n)_k (z^2)^k}{(\frac{1}{2})_k \, k!}$$

where we have used that $(2k)! = \left(\frac{1}{2}\right)_k 2^{2k} k!$. Hence, one obtains the relation as follows

$$H_{2n}(z) = (-1)^n \frac{(2n)!}{n!} F\left(-n, \frac{1}{2}; z^2\right).$$

Similarly,

$$H_{2n+1}(z) = (-1)^n \frac{(2n+1)!}{n!} 2z F\left(-n, \frac{3}{2}; z^2\right).$$

To express the Bessel functions via the degenerate hypergeometric one assume that $\mathrm{Re}\, \nu > -\frac{1}{2}$ and use the integral representation (5.9). Changing the variables $s = \frac{1}{2}(1+t)$ in it, we obtain the relation:

$$J_\nu(z) = \frac{2^{2\nu}\left(\frac{z}{2}\right)^\nu e^{-iz}}{\Gamma\left(\frac{1}{2}\right)\Gamma\left(\nu+\frac{1}{2}\right)} \int_0^1 e^{2izs} s^{\nu-1/2}(1-s)^{\nu-1/2}\, ds$$

$$= \frac{2^{2\nu}\left(\frac{z}{2}\right)^\nu e^{-iz}\Gamma\left(\nu+\frac{1}{2}\right)}{\Gamma\left(\frac{1}{2}\right)\Gamma\left(\nu+\frac{1}{2}\right)} F\left(\nu+\frac{1}{2}, 2\nu+1; 2iz\right),$$

which can be rewritten as:

$$J_\nu(z) = \frac{\left(\frac{z}{2}\right)^\nu}{\Gamma(\nu+1)} e^{-iz} F\left(\nu+\frac{1}{2}, 2\nu+1; 2iz\right), \quad |\arg z| < \pi,$$

which establishes the relation between the two special functions under consideration.

Chapter 9
θ-Functions

Abstract A part of this section is recorded by MIPT student Anton Piankov. It contains the derivation of various properties of the θ-functions. It also contains an elementary description of the relation of these functions to the Riemann geometry.

Consider the one dimensional Shrödinger equation:

$$-i\frac{\partial}{\partial \tau}\theta(z|\tau) = -\frac{\pi}{4}\frac{\partial^2}{\partial z^2}\theta(z|\tau). \tag{9.1}$$

Any free one dimensional Shrödinger equation can be transformed into such a form by a suitable rescaling of τ and z. Obviously $f_n(z|\tau) \equiv e^{n^2 \pi i \tau + 2niz}$ solves this equation. Hence, obviously also the following function solves it:

$$\theta(z|\tau) \equiv \sum_{n=-\infty}^{+\infty} e^{n^2 \pi i \tau + 2niz}.$$

Consider now complex τ and z. If $\operatorname{Im} \tau > 0$, then this series is convergent. Hence, $\theta(z|\tau)$ is an analytic function of z.

Defining

$$q \equiv e^{i\pi\tau},$$

we can rewrite the function under consideration as follows:

$$\theta(z, q) \equiv 1 + 2\sum_{n=1}^{+\infty} q^{n^2} \cos(2nz).$$

Then, it is easy to see that

$$\theta(z + \pi, q) = \theta(z, q).$$

V. Akhmedova and E. T. Akhmedov, *Selected Special Functions*
for Fundamental Physics, SpringerBriefs in Physics,
https://doi.org/10.1007/978-3-030-35089-5_9

Furthermore,

$$\theta(z + \pi\tau, q) = \sum_{n=-\infty}^{+\infty} q^{n^2} q^{2n} e^{2niz} = \frac{e^{-2iz}}{q} \sum_{n=-\infty}^{+\infty} q^{(n+1)^2} e^{2(n+1)iz},$$

and, hence,

$$\theta(z + \pi\tau, q) = \frac{e^{-2iz}}{q} \theta(z, q).$$

The function $\theta(z, q)$ is referred to as θ-function. It is quasi-doubly periodic function of z: 1 and e^{-2zi}/q are so called periodicity multipliers. I.e. $\theta(z, q)$ is almost a function on the two-dimensional Riemann torus.

This torus appears if we divide the complex z-plane by the group that acts as follows

$$(z \to z + \pi, z \to z + \pi\tau).$$

In fact, one obtains a two-dimensional torus by the factorization over this group as is shown on the figure:

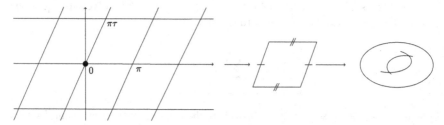

Under the action of such a group all parallelogram cells are identified to each other and their boundaries are identified according to this picture.

9.1   Different Types of θ-Functions

The function $\theta(z, q)$ defined above is usually denoted as $\theta_3(z, q)$. Similarly one can define functions as follows:

$$\theta_4(z, q) \equiv \theta_3\left(z - \frac{1}{2}\pi, q\right), \quad \text{i.e.} \quad \theta_4(z, q) = \sum_{n=-\infty}^{+\infty} (-1)^n q^{n^2} e^{2niz}$$

or

$$\theta_4(z, q) = 1 + 2 \sum_{n=-\infty}^{+\infty} (-1)^n q^{n^2} \cos(2\, n\, z).$$

Furthermore,

$$\theta_1(z, q) \equiv -i\, e^{iz + \frac{\pi i \tau}{4}} \theta_4\left(z + \frac{1}{2}\pi\tau, q\right) = -i \sum_{n=-\infty}^{+\infty} (-1)^n q^{(n+\frac{1}{2})^2} e^{(2n+1)z\, i},$$

and

$$\theta_1(z, q) = 2 \sum_{n=-\infty}^{+\infty} (-1)^n q^{(n+\frac{1}{2})^2} \sin[(2n+1)z].$$

Finally:

$$\theta_2(z, q) \equiv \theta_1\left(z + \frac{1}{2}\pi, q\right) = 2 \sum_{n=-\infty}^{+\infty} q^{(n+\frac{1}{2})^2} \cos[(2n+1)z].$$

Obviously, $\theta_1(z, q)$ is odd function of z, while $\theta_2(z, q), \theta_3(z, q)$ and $\theta_4(z, q)$ are all even. It is straightforward to see that all these functions solve the differential equation (9.1). Also all of them have the same periodicity multipliers 1 and $\frac{e^{-2iz}}{q}$.

Note that frequently instead of $\theta_i(z, q)$ they write $\theta_i(z)$, $i = \overline{1, 4}$ and instead of $\theta_i(0)$ they write θ_i, $i = \overline{1, 4}$. We will use these notations below.

9.2 Zeros of the θ-Functions

From the quasi-periodicity of $\theta_i(z)$ it is obvious that if z_0 is its zero, i.e. $\theta_i(z_0) = 0$, then $z_0 + m\pi + n\pi\tau$, for $n \in \mathbb{Z}$ and $m \in \mathbb{Z}$ is also its zero. Then consider cell parallelogram D in the complex z-plane, which is shown on the figure:

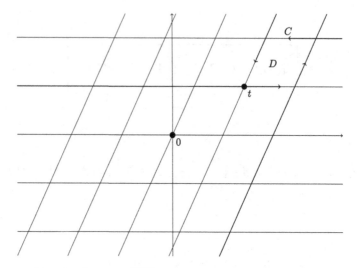

We now show that $\theta_i(z)$, $i = \overline{1,4}$ have one and only one zero inside each cell D. In fact, because $\theta_i(z)$ is analytic in any finite part of z-plane, hence, the number of zeros, n_0, of $\theta_i(z)$ inside a cell D is equal to

$$n_0 = \frac{1}{2\pi i} \oint_C \frac{\theta_i'(z)}{\theta_i(z)} dz,$$

where the contour C is the boundary of the cell D. One can choose any of the cells depicted on the figure above.

But

$$\frac{1}{2\pi i} \oint_C \frac{\theta_i'(z)}{\theta_i(z)} dz = \frac{1}{2\pi i} \left\{ \int_t^{t+\pi} + \int_{t+\pi}^{t+\pi+\pi\tau} + \int_{t+\pi+\pi\tau}^{t+\pi\tau} + \int_{t+\pi\tau}^{t} \right\} \frac{\theta_i'(z)}{\theta_i(z)} dz,$$

where t is the lower left angle of the cell D in the complex z-plane.

In the second and the third integrals one can make the following change of variables: $z + \pi$ and $z + \pi\tau$ for z. Then, we obtain:

$$\frac{1}{2\pi i} \oint_C \frac{\theta_i'(z)}{\theta_i(z)} dz = \frac{1}{2\pi i} \int_t^{t+\pi} \left[\frac{\theta_i'(z)}{\theta_i(z)} - \frac{\theta_i'(z+\pi\tau)}{\theta_i(z+\pi\tau)} \right] dz - \frac{1}{2\pi i} \int_t^{t+\pi\tau} \left[\frac{\theta_i'(z)}{\theta_i(z)} - \frac{\theta_i'(z+\pi)}{\theta_i(z+\pi)} \right] dz. \quad (9.2)$$

Now, if one uses that

$$\frac{\theta_i'(z+\pi)}{\theta_i(z+\pi)} = \frac{\theta_i'(z)}{\theta_i(z)}$$

and

$$\frac{\theta_i'(z + \pi\tau)}{\theta_i(z + \pi\tau)} = -2i + \frac{\theta_i'(z)}{\theta_i(z)},$$

he obtains:

$$\frac{1}{2\pi i} \oint_C \frac{\theta_i'(z)}{\theta_i(z)} dz = \frac{1}{2\pi i} \int_t^{t+\pi} 2i\,dz = 1.$$

Thus, $\theta_i(z)$ has only one zero in each cell. Obviously $z = 0$ is zero of $\theta_1(z)$. Hence, obviously zeros of $\theta_2(z)$, $\theta_3(z)$ and $\theta_4(z)$ are correspondingly at $z = \frac{1}{2}\pi$, $\frac{1}{2}\pi + \frac{1}{2}\pi\tau$ and $\frac{1}{2}\pi\tau$.

9.3 Composition Equations

Consider the following product $\theta_3(z + y)\theta_3(z - y)$ as a function of z. Then periodicity multipliers of this function corresponding to the periods π and $\pi\tau$ will be 1 and $\frac{e^{-2i(z+y)}}{q} \cdot \frac{e^{-2i(z-y)}}{q} = \frac{e^{-4iz}}{q}$. But the function $a\theta_3^2(z) + b\theta_1^2(z)$ has the same periodicity multipliers for the same periods. Hence, we can choose $\frac{a}{b}$ (a and b do not depend on z, but can depend on y) such that the doubly periodic function $\frac{a\theta_3^2(z)+b\theta_1^2(z)}{\theta_3(z+y)\theta_3(z-y)}$ will not have a pole at the position of the zero of the function $\theta_3(z - y)$.

Then, this function will have at most one simple pole in every cell D at the position of the zero of the function $\theta_3(z + y)$. As the result, this function will be constant, i.e. z-independent. Because for any doubly periodic function $f(z)$ it is true that $\frac{1}{2\pi i} \oint_C f(z) dz = 0$, where C is the boundary of the corresponding cell D. Hence, the sum of all residues of all doubly periodic functions should be 0. I.e. they can be either constant or have at least two poles inside each cell.

Thus, by adjusting the ratio $\frac{a}{b}$ we can choose this constant to be such that:

$$a\,\theta_3^2(z) + b\,\theta_1^2(z) = \theta_3(z + y)\,\theta_3(z - y).$$

To find a and b let us put $z = 0$ and then $z = \frac{\pi}{2} + \frac{\pi\tau}{2}$. As a result, we obtain the following relations:

$$a\theta_3^2 = \theta_3^2(y), \quad \text{and} \quad b\theta_1^2\left(\frac{\pi}{2} + \frac{\pi\tau}{2}\right) = \theta_3\left(\frac{\pi}{2} + \frac{\pi\tau}{2} + y\right)\theta_3\left(\frac{\pi}{2} + \frac{\pi\tau}{2} - y\right).$$

Hence, $a = \frac{\theta_3^2(y)}{\theta_3^2}$ and $b = \frac{\theta_1^2(y)}{\theta_3^2}$ and there is the following relation:

$$\boxed{\theta_3(z + y)\theta_3(z - y)\theta_3^2 = \theta_3^2(y)\theta_3^2(z) + \theta_1^2(y)\theta_1^2(z),}$$

which is one of the so called composition equations for the θ-functions.

In a similar manner one can find other relations between θ-functions. E.g.

$$\theta_2(y+z)\theta_2(y-z)\theta_4^2 = \theta_4^2(y)\theta_2^2(z) - \theta_1^2(y)\theta_3^2(z) = \theta_2^2(y)\theta_4^2(z) - \theta_3^2(y)\theta_1^2(z).$$

If we put $y = 0$ in the last expression we obtain the relation between squares of the θ-functions:

$$\theta_2^2(z)\theta_4^2 = \theta_4^2(z)\theta_2^2 - \theta_1^2(z)\theta_3^2.$$

Analogous relations follow from other similar equations.

9.4   Infinite Product Representation

Consider the following infinite product

$$f(z) = \prod_{n=1}^{\infty} \left(1 - q^{2n-1}e^{2iz}\right) \prod_{n=1}^{\infty} \left(1 - q^{2n-1}e^{-2iz}\right).$$

Because for $\mathrm{Im}\,\tau > 0$ the series $\sum_{n=1}^{\infty} q^{2n-1}$ is absolutely convergent, $f(z)$ is analytic for any finite part of the complex z-plane.

   The zeros of $f(z)$ are at

$$e^{2iz} = e^{(2n+1)\pi\tau i}, \quad \text{hence,} \quad 2zi = (2n+1)\pi\tau i + 2\pi m i, \quad n, m \in \mathbb{Z}.$$

Thus, $f(z)$ and $\theta_4(z)$ have the same zeros. As the result $\frac{\theta_4(z)}{f(z)}$ does not have neither zeros nor poles in any finite part of the z-plane.

   Furthermore, obviously, $f(z+\pi) = f(z)$. Also

$$f(z+\pi\tau) = \prod_{n=1}^{\infty} \left(1 - q^{2n+1}e^{2iz}\right) \prod_{n=1}^{\infty} \left(1 - q^{2n-3}e^{-2iz}\right)$$

$$= \frac{f(z)(1 - q^{-1}e^{-2iz})}{1 - qe^{2iz}} = -\frac{e^{-2iz}}{q}f(z).$$

As the result $\frac{\theta_4(z)}{f(z)}$ is doubly periodic without zeros and poles. Hence, it is constant. As the result we have the relation as follows:

$$\theta_4(z) = G \prod_{n=1}^{\infty} \left(1 - 2q^{2n-1}\cos(2z) + q^{4n-2}\right)$$

for some constant G which we will find in a moment.

Changing in this expression z for $z + \frac{\pi}{2}$, we obtain

$$\theta_3(z) = G \prod_{n=1}^{\infty} \left(1 + 2q^{2n-1} \cos(2z) + q^{4n-2}\right).$$

Moreover,

$$\theta_1(z) = -i\, q^{\frac{1}{4}}\, e^{iz}\, \theta_4 \left(z + \frac{\pi\tau}{2}\right)$$

$$= -i\, q^{\frac{1}{4}}\, e^{iz}\, G \prod_{n=1}^{\infty} \left(1 - q^{2n}\, e^{2zi}\right) \prod_{n=1}^{\infty} \left(1 - q^{2n-2}\, e^{-2zi}\right)$$

$$= 2\, G\, q^{\frac{1}{4}}\, \sin(z) \prod_{n=1}^{\infty} \left(1 - q^{2n}\, e^{2zi}\right) \prod_{n=1}^{\infty} \left(1 - q^{2n}\, e^{-2zi}\right).$$

Thus,

$$\theta_1(z) = 2\, G\, q^{\frac{1}{4}}\, \sin(z) \prod_{n=1}^{\infty} \left(1 - 2q^{2n} \cos(2z) + q^{4n}\right).$$

And finally,

$$\theta_2(z) = \theta_1 \left(z + \frac{\pi}{2}\right) = 2\, G\, q^{\frac{1}{4}}\, \cos(z) \prod_{n=1}^{\infty} \left(1 + 2q^{2n} \cos(2z) + q^{4n}\right).$$

To find G, let us represent $\theta_1(z)$ as $\theta_1(z) = \sin(z)\, \phi(z)$. Then:

$$\theta_1'(z) = \cos(z)\, \phi(z) + \sin(z)\, \phi'(z).$$

Hence, $\theta_1'(0) = \phi(0)$. At the same time:

$$\theta_1'(0) = 2\, G\, q^{\frac{1}{4}} \prod_{n=1}^{\infty} \left(1 + q^{2n}\right)^2, \quad \theta_2(0) = 2\, G\, q^{\frac{1}{4}} \prod_{n=1}^{\infty} \left(1 + q^{2n}\right)^{2n},$$

$$\theta_3(0) = G \prod_{n=1}^{\infty} \left(1 + q^{2n-1}\right)^2 \quad \text{and} \quad \theta_4(0) = G \prod_{n=1}^{\infty} \left(1 - q^{2n-1}\right)^2.$$

There is a relation stating that:

$$\theta_1' = \theta_2\theta_3\theta_4, \tag{9.3}$$

which we prove below. From this relation we obtain that:

$$\prod_{n=1}^{\infty} \left(1 - q^{2n}\right)^2 = G^2 \prod_{n=1}^{\infty} \left(1 + q^{2n}\right)^2 \left(1 + q^{2n-1}\right)^2 \left(1 - q^{2n+1}\right)^2.$$

Because $|q| < 1$ the whole product is absolutely convergent. Hence, it's true that:

$$\prod_{n=1}^{\infty} \left(1 - q^{2n-1}\right) \prod_{n=1}^{\infty} \left(1 - q^{2n}\right) \prod_{n=1}^{\infty} \left(1 + q^{2n-1}\right) \prod_{n=1}^{\infty} \left(1 + q^{2n}\right) =$$

$$= \prod_{n=1}^{\infty} \left(1 - q^n\right) \prod_{n=1}^{\infty} \left(1 + q^n\right) = \prod_{n=1}^{\infty} \left(1 - q^{2n}\right).$$

Therefore:

$$G^2 = \prod_{n=1}^{\infty} (1 - q^{2n})^2,$$

and, hence,

$$G = \pm \prod_{n=1}^{\infty} (1 - q^{2n}).$$

To define the sign, notice that for $|q| < 1$ the function G is analytic in q and, when $q \to 0$ we obtain that $G \to 1$, as can be seen from the product defining $\theta_3(z)$. Hence it is true that:

$$G = \prod_{n=1}^{\infty} (1 - q^{2n}).$$

To prove the relation (9.3), let us take the derivative of $\log \theta_3(z)$:

$$\theta_3'(z) = \theta_3(z) \left[\sum_{n=1}^{\infty} \frac{2i\,q^{2n-1}\,e^{2iz}}{1 + q^{2n-1}\,e^{2iz}} - \sum_{n=1}^{\infty} \frac{2i\,q^{2n-1}\,e^{-2iz}}{1 + q^{2n-1}\,e^{-2iz}} \right].$$

Then

$$\theta_3''(z) = \theta_3'(z) \left[\sum_{n=1}^{\infty} \frac{2\, i\, q^{2n-1}\, e^{2zi}}{1 + q^{2n-1}\, e^{2zi}} - \sum_{n=1}^{\infty} \frac{2\, i\, q^{2n-1}\, e^{-2zi}}{1 + q^{2n-1}\, e^{-2zi}} \right]$$

$$+ \theta_3(z) \left[\sum_{n=1}^{\infty} \frac{(2i)^2\, q^{2n-1}\, e^{2zi}}{\left(1 + q^{2n-1}\, e^{2zi}\right)^2} + \sum_{n=1}^{\infty} \frac{(2i)^2\, q^{2n-1}\, e^{-2zi}}{\left(1 + q^{2n-1}\, e^{-2zi}\right)^2} \right].$$

Now taking the limit $z \to 0$, we obtain that

$$\theta_3'(0) = 0 \quad \text{and} \quad \theta_3''(0) = -8\, \theta_3 \sum_{n=1}^{\infty} \frac{q^{2n-1}}{\left(1 + q^{2n-1}\right)^2}.$$

Performing similar manipulations with θ_4 and θ_2, we obtain that

$$\theta_4'(0) = 0 \quad \text{and} \quad \theta_4''(0) = 8\, \theta_4 \sum_{n=1}^{\infty} \frac{q^{2n-1}}{\left(1 - q^{2n-1}\right)^2}.$$

Similarly

$$\theta_2'(0) = 0 \quad \text{and} \quad \theta_2''(0) = \theta_2 \left[-1 - 8 \sum_{n=1}^{\infty} \frac{q^{2n}}{\left(1 + q^{2n}\right)^2} \right].$$

Using that $\theta_1(z) = \sin(z)\, \phi(z)$ one can find:

$$\phi'(0) = 0 \quad \text{and} \quad \phi''(0) = 8\, \phi(0) \sum_{n=1}^{\infty} \frac{q^{2n}}{\left(1 - q^{2n}\right)^2}.$$

Then $\theta_1'(0) = \phi(0)$ an $\theta_1'''(0) = 3\, \phi''(0) - \phi(0)$. From here we obtain that:

$$\frac{\theta_1'''(0)}{\theta_1'(0)} = 24 \sum_{n=1}^{\infty} \frac{q^{2n}}{\left(1 - q^{2n}\right)^2} - 1. \tag{9.4}$$

From all the found in this subsection relations together it follows that:

$$1 + \frac{\theta_2''(0)}{\theta_2(0)} + \frac{\theta_3''(0)}{\theta_3(0)} + \frac{\theta_4''(0)}{\theta_4(0)}$$

$$= 8 \left[-\sum_{n=1}^{\infty} \frac{q^{2n}}{\left(1 + q^{2n}\right)^2} - \sum_{n=1}^{\infty} \frac{q^{2n-1}}{\left(1 + q^{2n-1}\right)^2} + \sum_{n=1}^{\infty} \frac{q^{2n-1}}{\left(1 - q^{2n-1}\right)^2} \right]$$

$$= 8 \left[- \sum_{n=1}^{\infty} \frac{q^n}{(1 + q^n)^2} + \sum_{n=1}^{\infty} \frac{q^n}{(1 - q^n)^2} - \sum_{n=1}^{\infty} \frac{q^{2n}}{(1 - q^{2n})^2} \right].$$

To obtain the last equality we have joined the first two terms into the single term containing summation over all powers n rather than just over even or odd separately. Also we have represented the third term as the difference of two terms.

Finally, combining the first two series in the last expression, we obtain:

$$1 + \frac{\theta_2''(0)}{\theta_2(0)} + \frac{\theta_3''(0)}{\theta_3(0)} + \frac{\theta_4''(0)}{\theta_4(0)}$$

$$= 24 \sum_{n=1}^{\infty} \frac{q^{2n}}{\left(1 - q^{2n}\right)^2} = 1 + \frac{\theta_1'''(0)}{\theta_1'(0)},$$

where on the last step we have used the Eq. (9.4).

Using this equation and the Schrödinger equations that are obeyed by each of the θ-functions, we obtain that:

$$\frac{1}{\theta_1'(0|\tau)} \frac{d\theta_1'(0|\tau)}{d\tau} = \frac{1}{\theta_2(0|\tau)} \frac{d\theta_2(0|\tau)}{d\tau} + \frac{1}{\theta_3(0|\tau)} \frac{d\theta_3(0|\tau)}{d\tau} + \frac{1}{\theta_4(0|\tau)} \frac{d\theta_4(0|\tau)}{d\tau}.$$

Integrating this expression over τ, one can find that:

$$\theta_1'(0, q) = C\, \theta_2(0, q)\, \theta_3(0, q)\, \theta_4(0, q)$$

for some integration constant C, which is independent of q. To define this constant let us take the limit $q \to 0$ in the found expression. Because,

$$\lim_{q \to 0} q^{-\frac{1}{4}} \theta_1' = 2, \quad \lim_{q \to 0} q^{-\frac{1}{4}} \theta_2 = 2, \quad \lim_{q \to 0} \theta_3 = 1, \quad \text{and} \quad \lim_{q \to 0} \theta_4 = 1,$$

we find that $C = 1$ and, as the result, Eq. (9.3) follows.

Printed in the United States
By Bookmasters